BIG QUESTIONS
IN SCIENCE

Harriet Swain is the deputy features editor at the *Times Higher Education Supplement*.

John Maddox, editor emeritus of *Nature*, is a trained chemist, taught theoretical physics at the University of Manchester and was editor of the science journal *Nature* for 22 years. He has served on royal commissions on environmental pollution and genetic manipulation and as director of the Nuffield Foundation. His books include *The Spread of Nuclear Weapons* (with Leonard Beaton), *Revolution in Biology*, *The Doomsday Syndrome*, *Beyond the Energy Crisis* and *What Remains to be Discovered*.

BIG QUESTIONS IN SCIENCE

EDITED BY
Harriet Swain

INTRODUCTION BY
John Maddox

VINTAGE

Published by Vintage 2003

2 4 6 8 10 9 7 5 3 1

First published in Great Britain in 2002 by
Jonathan Cape

Vintage
Random House, 20 Vauxhall Bridge Road,
London SW1V 2SA

Random House Australia (Pty) Limited
20 Alfred Street, Milsons Point, Sydney
New South Wales 2061, Australia

Random House New Zealand Limited
18 Poland Road, Glenfield,
Auckland 10, New Zealand

Random House (Pty) Limited
Endulini, 5A Jubilee Road, Parktown 2193,
South Africa

The Random House Group Limited Reg. No. 954009
www.randomhouse.co.uk

A CIP catalogue record for this book
is available from the British Library

ISBN 0 09 942892 X

Printed and bound in Great Britain by
Cox & Wyman Limited, Reading, Berkshire

Contents

Preface

'A key word in the vocabulary of a scientist is wonder,' writes John Polkinghorne, former president of Queens' College, Cambridge. For him, it is 'the response evoked by the marvellous order of the physical world revealed to our inquiry'.

Wonder is a key word in most people's vocabulary; they just seem to have fewer opportunities than scientists to exercise it. This book is an attempt to give them that opportunity – in response to questions that almost all will have pondered from time to time.

They are really big questions: How did the world begin? What is life about? But they are questions to which scientists are attempting to find answers – and in some cases apparently coming close to finding them.

Scientists spend their lives asking questions. The problem is that often they are asking them from contexts the rest of us do not understand, and the answers they find can be highly controversial. Here, therefore, journalists have been asked to introduce each question by highlighting some of these underlying contexts and controversies. They describe answers found in the past and by other current researchers and discuss new areas of investigation.

Thanks to both the journalists and academic contributors for their work in tackling such huge subjects, to Ed FitzGerald, Sian Griffiths, Val Pearce, Tracey Tucker and Neil Turner, and to Poppy Hampson and Will Sulkin at Jonathan Cape.

Special thanks to Mandy Garner, THES features editor, who made many of the initial approaches to writers, and to Auriol Stevens, former THES editor, for enabling the time and support that made this book possible.

Harriet Swain

? Introduction

Like John Polkinghorne's Anglican institutions, science is a broad church. There are probably as many lists of twenty big science questions as there are interested people – not least because of the different expectations of science held by practitioners on the one hand and non-scientists on the other.

Notoriously, scientists' big questions seem narrow to other mortals. Charles Darwin cudgelled his brains about the shapes of the beaks of finches in the Galapagos Islands before founding the theory of evolution that has triumphed for the past 150 years. In 1900, Max Planck began by seeking an explanation for the change in the quality of radiation from heated objects – the warmth of a domestic radiator is perceptible but invisible, whereas molten steel glows red – and ended by laying the foundations of counter-intuitive quantum mechanics. Part of the excitement of science is that answers to small questions repeatedly provoke much bigger ones.

The twenty big questions occupying the pages that follow have the virtue of being questions in *everybody*'s mind. All of us, after all, want to know how the world came into being and how, in due course (if at all), it will end. And how did uniquely human language, which allows us to discuss all this, come about?

Many of the big questions are also practical. That is as it should be. If science pretends to tell us how the universe began, it would be disreputable if it had nothing illuminating to say about how and why we fall in love, or whether (and how) hunger and pain can be eliminated.

Yet practical solutions to such problems are often extraneous to science itself. So much is clear from Brian Heap's account of the world's food supply: physical shortages of food are less important than the tariff regimes maintained by the major food-producing countries and the frequent indifference of the governments of poor countries to the economic well-being of their small farmers.

In my opinion, the decades ahead will be also riven by ethical arguments about the application of scientific knowledge. For example, is Prozac an acceptable route to the pursuit of human happiness? And for how long will Mary Warnock's admirably cautious welcome for the notion of engineering the improvement of the human condition (page 151) survive impatient demands for speedier action?

The plan of the book, which involves addressing each question at two levels, demonstrates just how elusive the answers can be. The introductory journalistic accounts that set out to explain previous attempts to formulate – and even answer – questions, prove how many false starts there have usually been. These articles are followed by attempts to provide a definitive answer to the question concerned by people eminently qualified for the task. My guess is that most readers will be impressed by the modesty and the tentativeness with which this part of the task is done. Very little, it seems, is cut and dried.

Nobody should be disappointed if the answers to these big questions rarely ring clear. Doubt about the way ahead has been a feature of science since Copernicus put the Sun rather than the Earth at the centre of the universe five hundred years ago. All answers are provisional until tested by experience or rendered invalid by unexpected discoveries.

That is not to say that there is no such thing as scientific truth. When Newton in the seventeenth century inferred from his new 'law of gravitation' that the Earth must be a sphere, he was correct, but only to a first approximation. That the Earth is not accurately a sphere, but is flattened at the poles because of its rotation, was quickly seen to be consistent with Newton's more general theory of mechanics.

The fact that in 1917 Einstein put Newton's law of gravitation into a broader philosophical setting (amending it in the process) did not make a fool of Newton; it proved Newton's genius. In turn, the exceptions (such as the deflection of distant starlight by the eclipsed Sun) proved Einstein's rule and made generally intelligible the circumstances in which Newton's writ still runs.

This succession of corrections illustrates what constitutes progress in post-Copernican science. The more perceptive the questions, the more challenging the research programmes they engender and the more tantalising the answers that emerge. The big surprises will be the answers to questions we are not yet smart enough to ask. The scientific enterprise is an unfinished project and will remain so for the rest of time. While our understanding of the world is enormously enlarged since Copernicus, it has still hardly begun.

If the answers in science are not always clear, neither are the questions. To the first of the questions in this volume, 'Does God exist?', I do not believe we shall ever have a scientific answer. To be sure, I am a non-believer, even a dis-believer (if that is a stronger word), but I feel it is a matter of simple logic that science can have nothing to say, one way or the other, about the reality of a supernatural guiding spirit.

I have a more contentious confession to make: I am not convinced by the big bang. Indeed, I believe it to be a kind of fairy story. Martin Rees, perhaps the most distinguished Astronomer Royal since Sir William Herschel, makes the contrary case on page 18. It is a beguiling one. As an account of how the universe began, the big bang theory has three huge advantages: almost by definition, it gives an expanding universe, it provides a natural explanation for the presence in the universe of heavy hydrogen (deuterium) and helium, neither of which can be manufactured in stars, as have been all the other elements in the world around us, and it offers a natural explanation of the low-temperature radiation that fills every corner of the universe.

But there are snags. The simple expansion of the big bang provides a universe much less homogeneous – much more clumpy – than the one in which we seem to live. Moreover, space-time should be curved (in the sense of Einstein's General Theory of Relativity), whereas it appears to be flat. Those difficulties have been finessed by the ingenious theory of the inflationary universe to which Sir Martin refers, but that device begs too many questions for my taste.

I do not apologise for that scepticism, except to acknowledge that I have no better model of the universe to put forward. Nor are imperfect scientific theories unusual. The big bang is the universally used model of the universe into which cosmologists seek to fit new data. It will continue in that role unless the imperfections multiply or until somebody has a dazzling new idea. Much the same is true of the human genome and the issue, much discussed in the following pages, of the relative importance of nature and nurture in the development of human beings.

Nobody seriously argues that we are exclusively the products of our genes. Simple-minded 'genetic determinism' is a false ogre. As Michael Rutter points out, both geneticists and psychologists have much to say, but also much to learn, about the nature/nurture interplay. The immediate challenge, for geneticists, is to make genetic sense of diseases such as diabetes in which people inherit susceptibility through several genes, and not just one (as with haemophilia). There is also the task of understanding in detail how the activity of genes in cells is regulated, both by internal signals and by signals from outside. In themselves, these are not 'big questions', but people will devote their lives to

Big Questions in Science

them and, as always, some of them will yield surprising and subversive insights. Genetic manipulation, already almost an exact science for bacteria, will, for example, come into its own for plants and animals when people have learned how to unpack and then repack naturally occurring DNA.

The psychologists have a higher mountain to climb. Neuroscience, as we know it, is almost entirely the product of the twentieth century, and it is still unclear how the brain enables us to think – even to make simple decisions, let alone to imagine circumstances never before experienced. Yet somehow the brain is an integrated record of all our experience and learning – a veritable repository of nurture – perhaps, as Susan Greenfield suggests (page 53), embodied in the myriad connections between nerve cells in the head. Especially because the mechanism of the brain's influence on the rest of the body remains shadowy except for the influence of sex hormones on sexual behaviour, decades will go by before the arresting hypotheses of the evolutionary psychologists are as well grounded as atomic science.

That is no scandal, but an illustration of how far science is a continuing enterprise. This volume is less an answer to the twenty big questions than a guide to how they may in due course be settled or replaced by others. So what kind of answers may we expect in the future?

It seems to me a matter of only decades before we have an authentic history of the human race, from the divergence of the human line from that of the Great Apes about 4.5 million years ago, through *Homo erectus* (who could stand up) to *Homo sapiens*, whose use of language has probably made possible the occupation of the entire surface of the Earth in the past 125,000 years or less. That will be done by a combination of genetics (especially the genetics of human embryology) and of classical palaeoanthropology. Further ahead is the prospect that the evolutionary tree of life will be reconstructed in fine detail.

Will that tell us how life began on the surface of the Earth? Not by itself. There is ample reason to believe that the first self-replicating molecules must have been much smaller chemicals than those now implicated in the sustenance of modern life. The search for life on other planets (which for the time being is a search for other planets like our own) would, if successful, radically change our view of our place in the world. But it will be hampered until we know how life began on the surface of the Earth within 100 million years or so of 4,000 million years ago.

The big question of the origin of life is a much neglected field of enquiry. But the decades ahead are bound to be dominated by the deepening understanding of life *processes* now flowing in a torrent from laboratories across the world – less than half a century since the structure of DNA proved to be the literal secret of life.

The potential benefits of these discoveries, not only for medicine but for the management of the planet, are immense, as John Sulston, Brian Heap and Ronald Melzack describe. Such benefits are threatened only by our collective impatience (from which scientists are not immune) – just take the misplaced zeal with which companies have sought to market genetically modified crops without having adequately demonstrated the safety of their proposals, or, as Mary Warnock warns, the possibility that advances in genetics will attempt to make mortality redundant. There are also, of course, the familiar difficulties of demonstrating that the pursuit of happiness can be a collective enterprise, not one in which there are more losers than winners.

John Maddox

Big Questions in Science

Does God Exist ?

'I am unaware of any irreconcilable conflict between scientific knowledge about evolution and the idea of a creator God,' says Francis Collins, the man behind the US human genome programme. 'I am a geneticist, yet I believe in God.'

Collins is not alone. Over the past few decades, many acclaimed scientists have declared their belief in God *and* science. They include: Sir John Houghton, co-chair of the scientists' working group of the Intergovernmental Panel on Climate Change; John Polkinghorne, former president of Queens' College, Cambridge, and a particle physicist turned Anglican priest; Carl Feit, immunologist at Yeshiva University in New York and a Talmudic scholar; and Russell Stannard, a physics professor at the Open University and a reader in the Church of England.

Such views are not universally held. Richard Dawkins, Oxford University's professor of the public understanding of science, and a vocal atheist, is quick to dismiss religious belief. He has called anyone advocating a creator God 'scientifically illiterate', while terming religion 'a virus'. Others see religion and science as different paradigms, both legitimate, but unrelated.

Yet some 3,000 years after a single creator God was first mooted, in an age when science is seeking to understand and control man's genetic make-up, and when powerful telescopes allow us to look to the very heart of the big-bang origin of the universe, the notion of God still persists. It is a notion that sells popular-science books, exercises theists and atheists alike, and is as plagued by division today as ever.

Belief in a single, perfectly good creator of the universe can probably be traced to ancient Israel around 1000 BC, says Richard Swinburne, Oxford University's professor of the philosophy of the Christian religion. Almost every society about which we have knowledge appears to have had faith in some divine force. It seems people have always looked to the divine for answers to questions that found no explanation in their society.

From the outset, there have been challenges. But, according to David Wilkinson, an astrophysicist turned minister and a fellow in Christian apologetics at St John's College, Durham University, questions about the existence, rather than the nature, of a creator have largely been a feature of more modern debate.

Until the mid nineteenth century, science and religion went, for the most part, hand in hand in Western society. Scientists typically explained their motivation in religious terms and many leading scientists were themselves clergy. Even the Church's persecution of Galileo, one of the most regularly cited examples of conflict between science and religion, did not involve his denial of the existence of God, although it did throw into question man's place at the centre of the universe.

The scientific revolution of the seventeenth century – with its development of instrumentation including microscopes – allowed scientists to marvel at the wonders of nature, and hence God. According to Wilkinson, design theory – the idea that nature is so well designed and so beautiful that it cannot be so by chance and must be the work of God – can be traced back to Greek literature, although, he says, its 'flourishing came with the scientific revolution'.

By the eighteenth century, some scientists were beginning to oppose religion, and by the early nineteenth century, the idea that the natural world was a simple mirror of God's work was under attack. It was Charles Darwin's *On the Origin of Species*, published in 1859, which presented, for some, the ultimate challenge to a creator God. Darwin undermined several traditional arguments for belief in God, including design theory and the unique status of humans.

Darwin at one time intended to join the clergy and was mindful not to use science to attack religion. Even at the time, some in the Church were open to his ideas and able to incorporate his work into their beliefs. However, some scientists were less accommodating. Thomas Huxley's legendary debate with the Bishop of Oxford in 1860 pitted evolutionary science firmly against belief in God. Two books of the time – William Draper's 1874 *History of the Conflict Between Religion and Science* and Andrew Dickson White's 1896 *History of the Warfare of Science With Theology in Christendom* – were instrumental in creating a confrontational image that endures today.

According to John Brooke, professor of science and religion at Oxford University, the post-Darwin period was not a simple conflict between science and religion, but it did spark moves to professionalise and secularise science, separating the discipline and its practitioners – who until the 1850s had been required to be devout Christians – from theology.

By the early twentieth century, a large number of scientists no longer believed in God. A 1916 survey of US scientists found 60 per cent did not believe in or doubted God – a figure that the author predicted would increase with the spread of education. But despite this, and despite significant advances in scientific understanding – particularly in the fields of genetics and quantum theory, which some believe obviate the need for a creator God – a 1996 survey still found 40 per cent of US scientists believed in God.

In a time when man has the power to play with life itself, how can there be room for God? The fact that the universe exhibits just the right conditions to foster life is one pointer, says the pro-God corner. 'Recent scientific work on the fine-tuning of the universe has shown that the initial matter and the laws of nature have to have very special features indeed if organisms are to evolve,' says Swinburne. The fact that our universe has just the right features may be chance or an indicator of the existence of a very large number of universes. Or it could be divine influence.

The existence of fundamental laws about the behaviour of matter is also mooted as evidence of a creator God. 'This is quite extraordinary,' says Swinburne. 'I believe God has a reason. By matter behaving this way, it is not only beautiful, but allows finite beings like us to make a difference to the world and each other.'

Some say human cognitive abilities, which greatly exceed the demands imposed by evolutionary pressures and allow us to perceive the complexities of the universe, point to God. Others cite the inability of science, as yet, to account completely for the origin of life. Though most scientists back ideas of biological evolution, there is less agreement on how natural selection began. Collins, director of the National Human Genome Research Institute at the National Institutes of Health in the United States, describes himself as a 'theistic evolutionist'. 'If God decided to create humans who he could have a relationship with, why should he not use the mechanism of evolution to do this?' he asks. 'It's an elegant idea.'

That no scientific proof of God has been forthcoming does not deter these theists, who also cite the scriptures and the wealth of human religious experiences as reasons for their belief. They suggest that science may well not be able to detect something as subtle as God.

Others go further. Michael Behe, a biochemist at Lehigh University in the US, says Darwinian evolution is incapable of explaining everything that exists in the living world. Rather, he says, God's handiwork can be seen in certain 'irreducibly complex' parts of organisms that could not have evolved from simpler components. His arguments for intelligent design have been attacked by many scientists, particularly in the US where scientific creationism – which takes a literal view of the biblical creation story – is still a major force.

Other scientists, including Harvard palaeontologist Stephen Jay Gould, though not dismissing the possibility of God, view science and religion as two different realms, logically distinct and fully separate in styles of enquiry and goals. He argues that science asks objective 'how' questions, while religion asks 'why'. Gould emphasises the need for the individual to use both to build a rich view of life. 'Science and religion should be equal, mutually respecting partners, each the master of its own domain with each domain vital to human life in a different way,' he says.

Those with less faith in the divine argue that the space for God is shrinking as scientific understanding of the universe increases. Rather than accommodating science and God, they believe science to be the only reliable path to knowledge. Dawkins perceives irreconcilable differences between science and religion. Championing Darwinian evolution, he sees this alone as sufficient explanation for diversity of life. And his view of a universe without design or purpose dismisses those who seek answers to 'why are we are here' type questions.

Nobel Prize-winning physicist Steven Weinberg agrees that 'the more the universe seems comprehensible, the more it seems pointless'. He argues that the prevalence of evil and misery proves that there is no benevolent designer.

Julia Hinde
Freelance writer on science and education

Does God exist?

John Polkinghorne
*Former president of Queens' College, Cambridge,
and an Anglican priest*

? Belief in the existence of God, as defined by concepts that would be held in common by the three great monotheistic world faiths, Judaism, Christianity and Islam, means that reality makes total sense and that the key explanatory principle needed for understanding is the recognition that the world is the creation of a divine agent. This, in turn, implies four statements that need to be defended: there is a Mind behind the order of the world; a Purpose behind its unfolding history; the One thus revealed is worthy of worship; and God is the ground of an everlasting hope.

A key word in the vocabulary of a scientist is 'wonder', the response evoked by the marvellous order of the physical world revealed to our enquiry. For those of us who have worked in fundamental physics, this feeling is particularly intense, for it has been our experience that the laws of nature are always expressed in equations of unmistakable mathematical beauty. Is it just our luck that the universe is both rationally transparent and rationally beautiful – in other words, that science is both possible and deeply rewarding? Personally, I cannot think that this remarkable fact is simply a happy accident. That the world is shot through with signs of mind becomes intelligible if there is indeed the Mind of its creator behind cosmic order. The argument is not a knock-down one, but it is coherent and intellectually satisfying.

The universe as we know it started with the big bang about fifteen billion years ago. It was extremely simple, an almost uniform expanding ball of energy. It is now rich and diversely structured. In the coming into being of self-conscious humankind, the universe became aware of itself, the most astonishing development that we know about in all that fifteen-billion-year history. These facts alone might suggest that more was happening in that history than just one thing after another. Moreover, while it took eleven billion years for any form of terrestrial life to develop, there is a real sense in which the universe was pregnant with that life from the start.

The scientific insights collected together under the rubric of the anthropic principle tell us that development of carbon-based life was only possible because the laws of nature that define the physical fabric of the world took a specific, finely tuned form. For instance, a delicate balance between gravity and electromagnetism is necessary if stars such as the Sun are to be able to burn steadily for the billions of years required for fuelling the development of life on an encircling planet. The nuclear forces had to be just right if carbon and the other elements essential to life were to be formed in the interior furnaces of first-generation stars.

Again, one may ask whether anthropic fine-tuning was just an incredibly happy accident or the sign that our world is the one prize drawn from a vast bran tub of other unobservable universes, all with different laws of nature. More persuasive for the theist than either of these proposals is the understanding that the anthropically fruitful universe has been endowed by God with those finely tuned circumstances that have enabled its fertile history to express the divine creative Purpose.

Of course, that cosmic potential has been fulfilled through evolutionary processes. It has involved interplay between 'chance' (not meaningless randomness but contingent particularity) and 'necessity' (anthropically fine-tuned law). This shuffling exploration of possibility does not pose a problem for the theist. God's creation is not a divine puppet theatre, for the Creator is no cosmic tyrant. Divine benevolence towards creation implies that there will be due degrees of independence granted to creatures.

To use a phrase that the clergyman Charles Kingsley

coined soon after publication of Charles Darwin's *On the Origin of Species*, an evolving world, seen from a theological point of view, is one in which creatures are allowed 'to make themselves'. This is a greater good than a ready-made creation, just as free beings are of greater worth than perfectly programmed automata, but it has a necessary cost in terms of the blind alleys of evolutionary exploration.

Here, science provides believers with some modest help as they struggle with their greatest difficulty: the existence of so much evil and suffering in what is claimed to be the creation of a good and powerful God. New forms of life will come about through genetic mutations, but exactly the same biochemical processes also mean that other mutant cells become malignant. One cannot have the one without the other. There is cancer in the world, not because the Creator is indifferent or incompetent, but because it is the unavoidable cost of a creation allowed to make itself. I do not at all suppose that this is a complete answer to the difficulties of suffering, but it does show that the presence of disease is not gratuitous.

Consideration of the One who is worthy of worship turns the argument in the direction of the nature and existence of value. Just as science is to be defended as the investigation of the existing order of the world, so the theist will claim that human culture at its deepest level is discovery and not invention, a response to the nature of reality and not the construction of systems of human-generated meaning. This assertion raises a vast array of issues on which there is much disagreement and the argument cannot be pursued in any detail here. I wish simply to affirm my belief that profound moral principles – such as truth being better than lies, or that persons are always ends and never merely means – are neither disguised survival strategies nor useful social conventions, but discernments of reality. We have real moral knowledge. In fact, it seems to me that I know that love is better than hate as surely as I know anything.

If moral principles are not simply matters of expediency or of our individual choosing, from where do they come and how do they possess their authority? The theist will see them as intimations of the good and perfect will of God. Similarly, aesthetic delight is a dimension of true encounter with reality, which the believer will understand as a sharing in the Creator's joy in creation. The encounter with the sacred, to which the world faith traditions testify, is a meeting with the divine presence.

The coherence and persuasiveness of theism depends partly on acknowledging the rich and many-layered nature of the reality within which we live. The same event can be both an occurrence in the physical world, a carrier of beauty, a challenge to moral decision and an encounter with the sacred. For the religious believer, an occasion of worship will often have all these dimensions. Belief in God ties together these levels of experience, whose coincidence might otherwise seem fortuitous. God is worthy of worship because God is ultimately the ground of the good, the true and the beautiful.

Finally, what about hope? We know that we are going to die and cosmology tells us that the universe itself will eventually end either in collapse or decay. So what sense will it all make ultimately? Is cosmic history, after all, just a tale told by an idiot? I think that there is a deep intuition in the human heart to the contrary, a trust that in the end all shall be well. The Marxist philosopher, Max Horkheimer, expressed the longing that the murderer should not tri-umph over the innocent victim. The eternal faithfulness of God is the only possible undergirding of such a hope.

Such a reliance on God raises a whole host of further questions about the divine nature and whether the Creator is really interested in any way in individual creatures. Responding to these questions requires turning from theism in general to the nature of Jesus Christ and his res-urrection. There is no space left to defend these specific Christian claims (happy as I would be to do so), but raising the issues serves to remind us that almost all the billions who believe in the existence of God do so not in a detached philosophical sort of way, but from within the experience of a living faith community. In turn, this raises further questions about how the faith traditions relate to each other, with their common ground of encounter with the sacred and their strikingly different descriptions of what that encounter reveals. Here is a problem that theology has only recently begun to treat sufficiently seriously. It will be a dominant item on its agenda during the twenty-first century, and probably beyond.

Further reading

John Brooke: *Science and Religion: Some Historical Perspectives* (Cambridge University Press, 1991)

Greg Easterbrook: 'Science and God: A Warming Trend' (*Science*, vol. 277, August 1997)

Stephen Jay Gould: *Rocks of Ages: Science and Religion in the Fullness of Life* (Ballantine Publishing Group, 1999)

John Polkinghorne: *The Way the World Is* (Triangle, 1983), *Science and Providence* (SPCK, 1989), *Belief in God in an Age of Science* (Yale University Press, 1998), *Faith, Science and Understanding* (Yale University Press, 2000)

Richard Swinburne: *Is there a God?* (Oxford University Press, 1996).

How Did the Universe Begin **?**

In many cultures, the origin of the universe is traditionally explained through tales of the creation or separation of the solid Earth, the waters and the sky. The stories commonly feature a belief that there is some device or being that acts as an intermediary between Heaven and Earth. For example, they depict real or mythical people and animals in celestial constellations.

In the modern era, most people regard these tales only as colourful traditions, part of cultural history rather than of science. Some, however, have survived – such as the Christian creationist belief that the universe was created in a few days about 6,000 years ago.

The flaw with all early accounts of the origin of the universe is that they devote the bulk of their attention to the formation of the Earth and its living creatures while, in the last few hundred years, it has become apparent that the Earth is a minute object in a massive universe, of interest only because we live there. It does not make even a footnote to a footnote in modern cosmology.

The study of the origin of the universe has turned, in the life-time of contemporary adults, from a field ripe for speculation into a genuine experimental science where theories are constrained by a growing range of fundamental data. The main agent of this change has been the telescope.

Astronomers use telescopes to look deep into space, but, for a cosmologist, telescopes can also peer into time. Light from the nearest star beyond the Sun takes four years to get here, so looking at it involves looking four years into the past. But modern telescopes look so deep into space that they are seeing the universe in the earliest era of its history.

In addition, modern technology allows virtually the whole of the electromagnetic spectrum to be viewed – not merely the small fraction apparent to the human eye. This has led to many insights, the most startling of which was the discovery by Nobel Prize-winning physicists Arno Penzias and Robert Woodrow Wilson of cosmic back-ground radiation, thought to be the leftover heat from the big bang, which is observed at radio wavelengths. It is regarded as a hole-in-one proof of the big-bang theory of the origin of the universe by almost all cosmologists.

In addition, the detection by American astronomer Edwin Hubble, early in the twentieth century, of the expansion of the universe has proved key to our understanding of its origin. The insight that led to its discovery was the realisation that lines seen in the spectra of distant galaxies were caused by transitions in the electron energies of known atoms, but with their wavelengths altered by the Doppler effect (the effect by which the frequency of radiation emitted by an object appears to increase the closer it is to the observer) because of their movement away from us.

This was a profound insight, built on decades of work in laboratory physics, which made astrophysical spectroscopy feasible. But it also depended on painstaking work to determine the scale of the universe and the distances of galaxies. In recent years, we have found out more about the amount of matter in the universe, its age and the crucial early period of inflation. The fact that the universe may have expanded for a short period with incredible speed was first suggested by Alan Guth of the Massachusetts Institute of Technology and is now generally accepted by most, although not all, scientists. Knowledge about this very early period has been made possible by observational astronomy working in tandem with developments in relativity and quantum theory.

These subjects can often seem arcane and almost incomprehensible. But we should recall that one of the first insights of relativity was the definitive solution to one of the longest running questions in science. For millennia, a range of ideas had been developed to explain how the stars, including the Sun, shine. Only in the twentieth century did the discovery of nuclear fusion provide the answer. It also confirmed that stars could indeed shine for the billions of years the Earth and other objects in the solar system had been independently proven to exist. Scientists have used this knowledge base to illuminate other big questions about major aspects of the universe, including the creation and distribution of the chemical elements, most of which have been produced in the cores of stars.

In modern cosmology, the universe appears to have been roughly the same sort of place for most of its fifteen-billion-year existence. It is a place to which the rules of physics that we determine on Earth apply, and have predictive power. For example, the appearance of the solar system – Sun, planets and satellites, asteroids and comets – about 4.5 billion years ago is a comprehensible event which can be modelled accurately and compared with observations of other stars, around which families of planets, and the dust discs from which they might form, are being discovered. By contrast, thinking about the earliest period of the universe takes us into a world where our normal ideas about time and space are valueless.

But there is a way of developing ideas about matter and energy under the extreme conditions that prevailed in the very earliest moments of the universe. Much as a telescope allows us to look back billions of years into the past, so current and planned particle accelerators allow conditions in the very early universe to be replicated. The scientific community that uses them overlaps with the telescope-based cosmology community and a healthy band of theoreticians.

One of the principal challenges they face is the existence of 'dark matter', unseen by telescopes or other instruments, but whose gravitational effects suggest that it may far outweigh the comparatively familiar material of which the observable universe is made. Its lack of luminosity does not mean that it is unobservable. Astronomical approaches to the problem include proposals to look for small dark-matter objects hiding stars. At the same time, there have been proposals that attack the problem via a particle physics approach. For example, it is known that the universe is flooded by particles called neutrinos, which are emitted by stars and have long been generally thought to have no mass. The missing mass problem might be solved if they had even a tiny amount. It is also possible that some unknown type of material is responsible for the missing mass.

One candidate is the WIMP (weakly interacting massive particle). WIMPs are being sought by experiments in old mines and other sites far away from interference. They are inherently hard to detect (because they interact only weakly with other matter) and if discovered would be a major addition to our knowledge of the taxonomy of matter.

But questions raised by the example of missing mass serve only to show the power of the 'standard model' of the universe in which almost all scientists believe. It unites the particles and forces that make up the universe within a single set of relationships whose shape is now being finalised. Even the discovery that the majority of the universe, in mass terms, may consist of material never seen in a laboratory has not caused a fundamental rethink of its basic accuracy. Part of the reason for the apparent stability of the standard model is its wide evidence base. Astronomical and laboratory methods have produced mutually reinforcing results.

The next stage in this collaboration will involve a further step change in equipment and ambition. The Large Hadron Collider (LHC) under construction beneath Switzerland and France is intended to explore the possible existence of the Higgs Boson, which explains why matter has mass. It is one of the particles predicted in the standard model but has so far proved elusive. The extreme conditions within the LHC will be the closest yet created to those at the origin of the universe.

In space, a clutch of instruments will build on the successes of the 1990s, including the Microwave Anisotropy Probe (MAP), a United States mission to examine the fine structure of the cosmic microwave background. The data it yields will tell us about the first moments after the big bang and about the way in which large-scale structures, such as galaxies, arose. MAP is a space mission, but the new generation of ground-based telescopes is also penetrating deep into time and allowing objects from within the first billion years of the universe to be observed.

New telescopes at the Earth's surface and in space, especially the successor to the Hubble Space Telescope, currently called the New Generation Space Telescope, will allow the universe to be observed at ever earlier stages of its development.

The connections between cosmology, cosmogony and physics now being formed confirm each other so closely that it might be tempting to assume that they provide a complete picture that allows for no improvement. But anyone tempted to believe that this subject can be closed in a few years should recall the state of physics a century ago, just before the discovery of radioactivity, the photoelectric effect and relativity. At that stage, a high level of confidence that our knowledge was close to perfection was the precursor of an unprecedented era of unexpected and far-reaching discovery that altered our perception of the universe completely.

Martin Ince
Deputy editor of the Times Higher Education Supplement

How did the universe begin?

Sir Martin Rees
*Astronomer Royal and professor of
astronomy at Cambridge University*

Evidence that our universe had a hot dense beginning dates back to 1965, when radio astronomers discovered that space was not completely cold, but had a temperature about three degrees above absolute zero. Indeed, our entire universe is pervaded with microwaves, which are an afterglow of the hot dense initial state. Since 1965, this evidence has become firmer. As a result, extrapolating cosmic history back to a stage when the universe had been expanding for a few seconds now deserves to be taken as seriously as, for instance, what geologists or palaeontologists tell us about the early history of our Earth; their inferences are just as indirect and generally less quantitative.

In contrast, we are still groping for firm clues about what happened still earlier – in the first tiny fraction of a second – when the primordial material was squeezed to such extreme densities and pressures that experiments offer no firm guide. Nevertheless, within the past few years, cosmologists have, remarkably, developed a consensus about how the universe is now expanding, and what its future is likely to be.

In about five billion years, the Sun will die, and the Earth with it. We cannot predict what role life will by then have carved out for itself; it could have become extinct, or it could have achieved such dominance that it influences entire galaxies. Such speculations are the province of science fiction, but cannot be dismissed as absurd. After all, it has taken little more than one billion years – only a fifth of the Sun's remaining life – for the first multi-cellular organisms to evolve.

But what will happen in the even more remote future? The answer depends on how much the cosmic expansion is being decelerated. Although everything exerts a gravitational pull on everything else, calculations show that if all the atoms in the universe were spread uniformly through space, they would not, in the absence of an external force, exert enough gravitational pull to slow things down. This implies perpetual expansion. Galaxies 'feel' the gravitational pull of several times more material than we actually see, most of which is 'dark matter'. But even taking the dark matter into account, there is not enough force to bring expansion of the universe to a halt. It is therefore forecast to continue expanding. Galaxies will fade from view as they get ever further away and their stars exhaust their fuel.

Recently, evidence has emerged that expansion is not slowing, but accelerating. This implies that, on the cosmic scale, gravity is overwhelmed by some kind of repulsive force. Such a force was hypothesised by Einstein in 1917. At that time, astronomers only really knew about our galaxy – not until the 1920s did a consensus develop that Andromeda and similar spiral-appearance collections of stars known as 'spiral nebulae' were actually separate galaxies, each comparable to our own. It was therefore natural for Einstein to presume that the universe was static – neither expanding nor contracting. He found that a universe could not persist in a static state unless an extra force counteracted gravity.

The motivation for his hypothesis became irrelevant after 1929 when Edwin Hubble discovered that the universe was expanding, but that does not discredit it. On the contrary, empty space now seems anything but simple. All kinds of particles are latent in it; on an even tinier scale, it may be a seething tangle of strings. From our modern perspective, the puzzle is not why there should be cosmic repulsion, but why the energy and force latent in empty space is not much higher.

Now, a remarkable concordance has emerged between independent methods for measuring the contents of the universe. It seems that atoms provide only 4 per cent of the mass energy in the universe, dark matter contributes 20 to 30 per cent, and the rest is 'dark energy' latent in space. There seems nothing 'natural' about this particular mixture. So how did it arise, and why is the universe expanding in the way it is?

When our universe was an amorphous fireball, only a second old, it could be described by just a few numbers: the proportions of ordinary atoms, dark matter and radiation, the expansion rate, and so forth. This simple recipe must be the outcome of what happened earlier still, within the first tiny fraction of a second, when conditions become more extreme and unfamiliar. In the first trillionth of a second, each particle would have carried more energy than can be reached by the most powerful accelerators at CERN, the European Laboratory for Particle Physics in Geneva. Ideas about this ultra early era are still tentative, but nonetheless there has been immense progress.

The most basic mystery is why our universe is expanding, and why it is so vast. Analogies with an explosion can be seriously misleading. Bombs on Earth, or supernovae in the cosmos, explode because a sudden boost in internal pressure flings the ejecta into a low-pressure environment. But in the early universe the pressure was the same everywhere; there was no empty region outside. The most plausible answers involve a so-called inflationary phase, during which the expansion was exponential; the scale doubled, then doubled, and then doubled again.

Within about 10^{-36} seconds, it is claimed, an embryo universe could have inflated enough to encompass everything we now see. The generic idea that our universe inflated from something microscopic is compellingly attractive and also accounts for why it is expanding. It looks like 'something for nothing', but that is not really the case. This is because our present vast universe may, in a sense, have zero net energy. Every atom has an energy because of its mass – Einstein's mc^2. But it also has a negative energy due to the gravitational field of everything

Big Questions in Science

else. So it does not, as it were, cost anything to expand the mass and energy in our universe.

Inflation stretches a microscopic patch until it becomes large enough to evolve into our observable universe. Indeed, it is likely to overshoot, inflating more than necessary. Our universe then ends up being 'stretched flat', rather like a part of a wrinkled surface becomes smooth if it is stretched enough.

Most theorists regard inflation as a beautiful generic concept, which they will cling to until something better comes along, and there are intimations that something might; extra spatial dimensions beyond the usual three may lead us to another paradigm, but the details depend on uncertain physics.

Observing some features of our present universe may help to select among rival theories. For example, inflation suggests an origin for the 'ripples' that show up as non-uniformities in the background temperature over the sky and are the embryos of galaxies. They are quantum vibrations, generated on a microscopic scale, that have inflated so much that they now stretch across the sky – an amazing link between cosmos and microworld. Some variants of the inflationary universe suggest that our big bang was not the only one – a speculation that dramatically enlarges our concept of reality and turns the history of our universe into just an episode, one facet, of the infinite multiverse. Astronomers are normally mere users of laboratory physics, except where gravity is concerned. Perhaps they can now return the compliment by probing 'extreme physics' that cannot be checked in the laboratory.

The crescendo of discovery seems set to continue throughout the present decade. Large telescopes can now view objects so far away that their light set out when the universe was only a tenth its present age. Other techniques can probe back to the first few seconds of the big bang. I would bet reasonable odds that within ten years we will know what the dominant dark matter is, and other key numbers like the age of the universe. If that happens, it will signal a great triumph for cosmology: we will have taken the measure of our universe, just as, over the last few centuries, we have learned the size and shape of our Earth and Sun.

In the longer term, theorists must elucidate the exotic physics of the very earliest stages. The ultimate synthesis that still eludes us is between gravity and the microworld – between the cosmos and the quantum. Until there is a unified theory, we will be unable to understand the fundamental features of our universe, imprinted at the very beginning, when everything was so squeezed that quantum fluctuations could shake the entire universe.

The smart money is on superstrings, or M theory, according to which each point in our ordinary three-dimensional space is actually a tightly folded origami in six or seven extra dimensions. There is still an unbridged gap between this elaborate mathematical theory and anything we can measure, but such a theory may be needed before we understand the beginning, or the nature, of the energy latent in empty space.

Further reading

John Barrow (with Frank Tipler, John Wheeler):
    The Anthropic Cosmological Principle (Oxford
    University Press 1988)
Roger Freedman and William Kaufmann: *Universe*
    (W.H. Freeman, 1999)
Alan Guth: *The Inflationary Universe* (Vintage,
    1998)
Stephen Hawking: *A Brief History of Time*
    (Bantam, 1998), *The Universe in a Nutshell*
    (Bantam, 2001)
Martin Rees: *Before the Beginning: Our Universe
    and Others* (Helix Books/Addison Wesley
    Longman, 1997), *Just Six Numbers: the Deep
    Forces that Shape the Universe* (Weidenfeld &
    Nicolson, 1999), *Our Cosmic Habitat*
    (Princeton University Press, 2001)
Steven Weinberg: *The First Three Minutes* (Basic
    Books, 1993).

What is Time?

? What is time? 'If nobody asks me,' St Augustine wrote in the fourth century, 'then I know; but if I were to desire to explain it to one that should ask me, plainly I know not.' Some sixteen centuries later the question remains as elusive as ever. Why does time seem to flow like a river, and what is the source of this river? The American physicist John Wheeler once suggested that 'time is what keeps everything from happening at once', a curiously seductive formula, though perhaps no less puzzling than the original question.

In his *Critique of Pure Reason*, the German philosopher Immanuel Kant argued that one could never perceive of or imagine anything existing outside space or in the absence of time. These are the 'subjective conditions of sensibility', he wrote. Much as a prism resolves light into its separate colours, laying them out in order, so does the mind, according to Kant, separate reality along the axis of time. But is time really only an illusion or a result of perception? Did it not exist long before there were any living things and so before there were any perceptions? Today, modern physics traces time's character back to the very origins of the universe and questions its place in the fundamental laws of physics.

Isaac Newton's equations of motion involve time but in a some-what sterile way. As the Earth orbits perpetually about the Sun, gravitation dictates a calculable change in the Earth's motion in each small interval of time. But this sort of time is merely a bookkeeper's trick, an artifice of accounting. For Newton, both space and time were absolutes: space a thoroughly empty void through which objects can pass, and time a kind of ticker tape running inexorably in the background. Albert Einstein then revealed that time could be stretched and distorted and was affected by matter and energy.

But even if the river of time flows faster in some places than others, and slows down when passing obstacles, this still does not explain what time is or why it has a direction. And experience suggests that time does have a direction. Washing machines wear out with use, as do automobiles and shoes, and they never return to their former pristine perfection. Mountain peaks crumble into the valley, but never reassemble themselves, and perfume from an opened bottle escapes to fill the room, but never does the reverse. These facts suggest a single direction for time, the direction in which things wear out, spread and erode away, and in which order generally dissolves into disorder.

Big Questions in Science

This tendency also points to a theoretical conundrum. The physics of perfume bottles, mountains and other large-scale things ought to arise out of the workings of their atoms and molecules. But, in contrast to the world around us, the atomic realm seems to make no distinction between past and future. Make a movie of a few atoms doing their thing, run it in reverse, and you would see nothing strange – the backward movement would again fulfil the laws of physics. But a movie of scattered rocks miraculously gathering again into a rugged peak would fly in the face of reality as we know it.

So how can the directionless time of the atomic realm give rise to the arrow of time at the larger scale? This is the central question, and its answer has two parts – the first relatively 'easy' and more than a century old, the second rather more difficult, a matter for continuing debate.

Why does perfume escape from a bottle, but never, on its own, re-enter? In the late nineteenth century, the Austrian physicist Ludwig Boltzmann reasoned as follows: suppose you calculate how many ways you can arrange a large number of perfume molecules so that they are spread out more or less uniformly through the room. Next, calculate how many ways you can arrange the same number of molecules while leaving them packed in the bottle. The former number, Boltzmann proved, is overwhelmingly huge when compared to the latter – bigger by a factor of 1 followed by more zeros than would fill every book in the British Library.

Now, the perfume molecules are slamming into one another, and generally flitting about randomly from one detailed arrangement to the next. And it follows, Boltzmann argued, that the perfume, unless somehow prevented, will tend to go from being packed neatly in the bottle to being scattered about outside of it. It is all down to the staggering mismatch in the number of ways the two situations can be realised.

No matter what you consider, there are always far more ways for the perfume's parts to be arranged in a disorderly way than in an orderly way. Disorder has a huge numbers advantage over order, and, as a result, things in our universe have a natural tendency to drift towards a condition of low-grade chaos. This is the Second Law of Thermodynamics: in the absence of any separate, organising force, things tend to drift in the direction of greater disorder, or greater 'entropy'.

Boltzmann's way of thinking offers the first striking insight into the nature of time – for it suggests that our subjective feeling of time is intimately wrapped up with the tendency things have to get muddled up and disorganised. The great German physicist Erwin Schrödinger once said: 'No perception in physics has ever seemed more important to me than that of Boltzmann.' The flow from order towards disorder seems to be a one-way stream, and this is why we sense a consistent direction in time, placing the unbroken wine glass or the brand new shoes in the past relative to their shattered or worn-out descendants.

But Boltzmann's great perception only brings into focus the deeper nub of the problem. A universal tendency for order to devolve towards disorder could explain why time seems to have a direction. But the explanation works only if we can explain how the universe got to be ordered in the first place. It could have started in a mess, dispersed like the perfume that has already spread through the room. Then there would be no gradual drift towards further disorder, and no direction to time. Explaining the direction of time means accounting for the great organisation initially present in the universe.

This is where many scientists are now focusing. In the perfume analogy, the stuff of our own universe was 'in the bottle' between ten and fifteen billion years ago, soon after the big bang. At that time, the distribution of energy and matter in the universe was extraordinarily smooth. In the early 1990s, physicists using the Cosmic Microwave Background Explorer telescope discovered just how smooth by studying the faint glow of microwave radiation that fills the universe and offers a snapshot of what it was like just 300,000 years after the big bang. They found that the distribution of matter then was uniform to 1 part in 100,000.

These observations set strong constraints on theories of the early universe. Of all the ways the stuff of the early universe could have been arranged, only a miniscule fraction would have given the universe the smoothness that astronomers' telescopes said it had. So the world was in a remarkably special condition – penned up and well prepared to let time loose. But how did it get that way?

The popular explanation is that the early universe experienced a special 'inflationary phase', a short-lived period in which the universe expanded with incredible speed, during which almost all ripples in the distribution of stuff would have been quickly ironed out. The idea of inflation makes the special smoothness not so special after all. It has gained support from measurements of the cosmic microwave background made by a telescope that flew over Antarctica by balloon in 1998. This telescope revealed ripples in the distribution of matter in the early universe of just the sort predicted by the inflationary idea.

Not everyone accepts the idea, however, or believes that it offers a full explanation. For such a theory could predict a bizarre future – such as a universe which eventually collapses on itself to produce a big crunch, during which time may reverse itself and flow backwards as things become more ordered rather than less, violating the Second Law of Thermodynamics. As Oxford mathematician Roger Penrose has pointed out, black holes, which result from the gravitational collapse of large stars, offer small-scale versions of such big crunch events, and yet work in a way that is fully in keeping with thermodynamics.

So our intuitive wondering about the nature of time has landed at the doorstep of the deepest issues in contemporary cosmology. In the fifth century BC, the Greek philosopher Parmenides went so far as to put all matters of time down to illusion, the true reality being eternal and unchanging. Some physicists and philosophers today might agree. Illusory or not, however, time's deepest secret has yet to be exposed.

Mark Buchanan
Physicist and science writer

What is time?

John Barrow

*Research professor of mathematical sciences and director of the
Millennium Mathematics Project at Cambridge University*

? Until the early years of the twentieth century, time was
seen as an inflexible ticking clock against which all
activity could be calibrated. Nothing could alter
time's metronomic course and its rate of passing was the
same everywhere for everyone. 'Now' was an unambiguous
notion, universally shared, and any changes in the rate of
flow of time were deemed to be subjective, as when Henry
Twells tells us:

> For when I was a babe and wept and slept,
>   Time crept;
> When I was a boy and laughed and talked,
>   Time walked;
> Then when the years saw me a man, Time ran;
> But as I older grew, Time flew.

Albert Einstein revealed concepts of space and time
that were vastly more complicated and mysterious than
any fiction writer had imagined. The flow of time was
determined by the mass and energy within it. As a result,
there was no absolute time, no unambiguous notion of
'now' for different observers, and no possibility of different
observers agreeing when events were simultaneous. There
were striking consequences. Time passes more slowly in
strong gravitational fields and for moving observers. Identical twins, sent on different space trips, will have different
ages on their return. All these counter-intuitive temporal
effects are routinely observed in the universe and deeply
woven into its tapestry of self-consistent symmetrical
laws.

But is time truly fundamental? It could be a simplifying approximate notion that emerges only in environments with low energies and temperatures – a classical limit of quantum reality. Such cool conditions are essential if atoms, molecules, and you and I are to exist. Yet, in the first moments of the universe's big-bang expansion, when energies and temperatures were vastly greater, a fundamentally different relationship between past and future could have existed, in which time was more like space.

We clock time in units that are astronomical in origin but anthropomorphic (days, months and years). Yet there is a superhuman unit of time that is defined solely by the laws and constants of nature. It is amazingly brief, just 10^{-43} of a second in duration. By this superhuman clock, our universe is about 10^{60} 'ticks' old. Strikingly, an age of at least 10^{59} is needed to produce the chemical elements needed for complexity and life. It is when the universe was only a few of these time units old that we suspect something exotically quantum may have happened to the nature of time. This is the realm in which gravity, relativity and quantum uncertainty amalgamate as equal partners to shape the universe.

The prime candidate for explaining such a partnership is string theory, known as M(ystery) theory. It predicts that there are more dimensions to space and time than the three and one we know. Physicists have generally assumed that all these extra dimensions are space-like, and all bar three are imperceptibly small, making it a challenge to detect their effects. But what if the extra dimensions are not all space-like? Some may be extra dimensions of time. What would that mean? What phenomena would it permit? Would it make particles decay too rapidly, or prevent 'observers' ever existing?

Extra dimensions of space open the door to changing the unchangeable. We believe there to be a distinguished collection of timeless building blocks that anchor the fabric of physical reality. We call them the 'constants of nature' and seek to measure them with greater precision and explain why they take the special numerical values that they do. In the former quest, we have been moderately successful; in the latter, we have drawn a complete blank. Nobody knows why the fundamental constants of nature take the numerical values that they do, only that if many of them were very slightly changed then neither we nor any other complex beings would be here to talk about it. But if there are more dimensions than three, then the true constants of nature are defined in all those dimensions and the 'shadows' of them that we see in our three-dimensional laboratories will change if the extra dimensions undergo any change in size.

Already there are tantalising hints that some of the observed 'constants' may not be constant when scrutinised at the limits of experimental precision. In the future, there will be special interest in these constants of nature and how they manage to resist the influence of time. Tests can be performed by monitoring high-precision atomic clocks on Earth over years or, more sensitively, by comparing the detailed atomic spectra from distant astronomical objects with the spectra from the same atoms on Earth. The light from the atomic transitions in material around distant quasars has taken nearly thirteen billion years to arrive at our telescopes. It is a time capsule telling us what physics was like far away and long ago when that spectral light began its journey across the universe.

How old is time? Until the mid 1970s, cosmologists believed that they had a powerful proof that once upon a time there was no time. A powerful mathematical theorem, developed by Cambridge theoretical physicist Stephen Hawking and Oxford mathematician Roger Penrose, showed that the attractive nature of gravity meant that the past must be finite.

In 1980, things began to change. Particle physicists found that their new theories abounded with possible forms of matter that possessed a tension that made them interact with themselves as if they were gravitationally repulsive. Suddenly, the proofs of a beginning to time were of mathematical interest only. There was no reason to believe that the assumptions on which they rested held good in the past.

We have found many appealing consequences of these gravitationally repulsive forms of matter. They could have made the universe accelerate very rapidly in its first moments of expansion and so explain how it came to be so big, so old, and so uniform, with just a smattering of lumps and bumps that subsequently provided the raw material for galaxies and stars. So far, the detailed predictions of this 'inflationary' universe theory agree well with observed patterns of radiation in the universe. Testing this theory with unprecedented precision was an aim of the Microwave Anisotropy Probe (MAP), a NASA space mission launched in July 2001.

Beyond the powers of MAP, unfortunately, is a fantastic by-product of the theory – one that complicates our picture of time to a potentially infinite degree. It suggests that the accelerated expansion of small parts of the universe will continue ad infinitum in a self-reproducing process that need have neither beginning nor end. Some of these bouts of fast expansion will create regions looking like our own visible universe, while others will produce conditions beyond our horizon, where even the number of large dimensions of space and time may be different.

Is time travel possible? Einstein was shocked when Austrian-born US philosopher and mathematician Kurt Gödel discovered that his theory of space, time and gravity permitted time travel. Maybe the universe was not safe for historians after all. But time travel in principle does not necessarily mean time travel in practice. The laws of physics allow fragments of broken glass to come together to create a beautiful wine glass, yet we never see such a thing because the motions required of the glass fragments are too improbable to be realised. Similarly, time travel may require unrealistically improbable conditions to be realised, even microscopically. Our best hope might be to find a measurable quantum process whose experimental outcome is significantly dependent upon the existence of time-travelling paths of information transfer. On the other hand, if time travel is possible, why do we not see evidence of it? Perhaps its consequences are always fatal, or maybe it requires a level of technical sophistication that no civilisation ever achieves because they all self-destruct, suffer cataclysmic impacts from space, or run out of the resources they need for technology. Or perhaps it is just too expensive. If its cost is minimal, then the most intriguing argument against its present occurrence is the existence of non-zero interest rates in the money markets. Only if interest rates are zero can time travellers not make huge gains by engaging in arbitrage transactions. And if they do make such profits, they will drive interest rates to zero!

Does time have a future? The immediate problem for cosmologists is to pin down how much matter there is in the universe, and determine whether its expansion has recently begun to accelerate, as observations are showing. As these observations are refined by increased accuracy, they will show how much time our asymptotic descendants will have to endure. Our universe does not appear to be expanding slowly enough to contract back to a big crunch. It seems destined to keep on expanding for ever. But while the universe may go on for ever, its constituents have a more limited life expectancy. Planets and stars will dismember and die; matter will decay; black holes will gorge themselves upon the debris and slowly evaporate, producing a dark and lonely future, populated by radiation and simple elementary particles. Life, perhaps more ethereal, disembodied and nanoscopic, will have to reform itself if it wishes to survive in any shape or form. If the universe has indeed recently embarked upon an unending phase of accelerated expansion, then even abstract 'life' has a limited future. Perhaps the perfection of backward time travel is the only long-term hope for materialists. Time is terribly patient.

Further reading

John Barrow: *The Anthropic Cosmological Principle*
(Oxford University Press, 1988), *Impossibility:*

the limits of science and the science of limits
(Oxford University Press, 1999), *Between Inner
Space and Outer Space* (Oxford University
Press, 2000), *The Book of Nothing* (Jonathan
Cape, 2000)

Roger Penrose: *The Emperor's New Mind* (Oxford
University Press, 1989)

Huw Price: *Time's Arrow and Archimedes' Point*
(Oxford University Press, 1996).

What is Consciousness ?

? Throughout the 1990s – christened the 'Decade of the Brain' by former US president George Bush – the scientific theory of consciousness was commonly called the last remaining challenge, the 'final frontier' of knowledge. Suddenly consciousness studies was back on the academic agenda after an embargo lasting nearly a century. According to philosopher John Searle, this embargo was a legacy of behaviourism, which asserted that scientific psychology should restrict itself to the study of observable phenomena, rather than subjective mental states. For a time it was considered such bad taste to raise the topic of consciousness in cognitive science discussions that graduate students would apparently 'roll their eyes at the ceiling and assume expressions of mild disgust'. So why the big change during the 1990s?

Some argue that consciousness studies regained academic respectability when Nobel laureates Francis Crick and Gerald Edelman started to work in the field in the 1980s. Not so, claims Semir Zeki of University College London, a leading pioneer of cognitive neuroscience. He puts it down to developments in brain-imaging techniques that enable psychologists to study the neural correlates of consciousness in glorious technicolour. But such was the stranglehold of behaviourism and its refusal to acknowledge anything other than the directly observable that even if this technology had been available a few decades earlier the dreaded C-word would still have been embargoed. Perhaps sociology offers a better explanation for its renaissance. It was, after all, during the 1990s that students educated in the 1960s, many of whom indulged a personal interest in altered states of consciousness, began to take up senior positions in cognitive science departments.

Although it is common to talk about *the* problem of consciousness, there are in fact three separate problems and differing views about how far they overlap. The technical term for the first problem of consciousness – described by Australian philosopher David Chalmers as the 'hard problem' – is the 'generation problem' – how do material configurations or processes produce conscious experience? In 1866, biologist T.H. Huxley expressed the problem as follows: 'How it is that anything so remarkable as a state of consciousness comes about as a result of irritating nervous tissue, is just as unaccountable as the appearance of the Djin when Aladdin rubbed his lamp.'

Since then, according to philosopher Jerry Fodor, we have made little progress: 'Nobody has the slightest idea how anything material could be conscious,' he says. 'Nobody even knows what it would be like to have the slightest idea about how anything material could be conscious. So much for the philosophy of consciousness.'

Few philosophers draw the conclusion that we should waste no more time on the problem, however. Fodor's colleague at Rutgers University, Colin McGinn, would agree that it is insoluble in principle. But he still manages to spend most of his time writing about it. Other 'mysterians', such as Oxford mathematician Roger Penrose or philosopher Bill Seager at the University of Toronto, resort to increasingly desperate measures – introducing everything from quantum mechanics to idealism (the theory that only the mental is real) and panpsychism (the doctrine that everything has mental properties) – to try to solve the problem through the 'principle of minimum mysteries' (quantum mechanics is a mystery, consciousness is a mystery, so perhaps they are one and the same mystery).

The fact that consciousness is a mystery is about the only thing on which Fodor agrees with Massachusetts Institute of Technology psychologist Steven Pinker, author of *How the Mind Works* (to which Fodor responded *The Mind Doesn't Work that Way*). At Tufts University, however, Daniel Dennett, another prominent advocate of evolutionary theory, thinks differently. According to Dennett, if we can provide an adequate explanation for cognitive abilities in functional, neurological and evolutionary terms then there is no 'further' problem of consciousness.

Neurophilosophers Paul and Patricia Churchland at the University of California, San Diego, broadly support this view, claiming that in the eyes of posterity the 'problem of consciousness' will look not unlike the 'problem of life'. They say that consciousness mysterians are the modern-day equivalent of vitalists, who believed life could not be explained purely by applying principles of chemistry and physics because it depended on a non-physical inner force.

Perhaps the best chance of resolving this dispute lies with a group of scholars who claim that the generation problem of consciousness is a product of the dualistic notion of mind as disembodied and distinct from its physical character, as expounded by mathematician and philosopher René Descartes (1596–1650). Nicholas Humphrey, a senior research fellow in the Centre for Philosophy of the Natural and Social Sciences at the London School of Economics, sees awareness as simply an activity of the nervous system. For him, sentience is just an evolved form of such activity, something that enacts rather than perceives. Its primitive origins are therefore observable even in the wriggles of single-cell organisms.

Other writers have drawn parallels with theories drawn from phenomenology, a tradition of Continental philosophy that studies consciousness from a first-person perspective. Cognitive neuroscientist Francisco Varela of the CNRS, the French National Research Centre, suggests neuroscience, phenomenology and behavioural science would have to unite to crack the problem of consciousness. Yet another group of writers, including neurobiologist Walter Freeman at the University of California, Berkeley, and philosopher Andy Clark, from Washington University, St Louis, draw on ideas of emergence and chaos theory to dissolve the mind/body problem. These suggest that the properties of consciousness emerge through minute fluctuations in the elements from which it is made and are therefore impossible to predict from those elements.

The second, related, problem of consciousness is the problem of the self. Modern debate on the problem of the self, although owing much to the insights of seventeenth- and eighteenth-century philosophers John Locke and David Hume, can still be contextualised within two schools of ancient thought. According to the ancient Greek philosopher Aristotle, who 'I' am is closely tied to my embodied existence. The psyche and the body form a unity and the psyche involves life functions from locomotion to philosophical contemplation. The opposing view, put by Pythagorean and Platonist thinkers, separated the psyche and the body and led directly to Descartes's theory of a disembodied ego.

In his *Essay concerning Human Understanding*, Locke addressed the problem of personal identity by proposing memory as the mechanism responsible for continuity of the self over time. When Hume turned his attention to this 'self', he found nothing more than a bundle of impressions with no introspective evidence that it existed. He therefore dismissed the self as a fiction of the imagination.

Modern philosophy of mind has done little more than refine this observation. Philosopher Anthony Kenny maintains that the self is a 'piece of philosophers' nonsense consisting in a misunderstanding of the reflexive pronoun', while Robert Nozick, also a philosopher, agrees that the self is 'synthesized in the act of reflexive self-reference' – in other words, it is simply a product of reflection.

Dennett's view of the self as a 'centre of narrative gravity' – a theorist's fiction which nonetheless plays a useful explanatory role – has been lent support by work on split-brain patients. Research by cognitive neuroscientist Michael Gazzaniga at Dartmouth College on these patients shows how the concept of the self is constructed by the 'interpreter' mechanism located in the left cerebral hemisphere. This function, which can often be dissociated in patients with brain damage, constructs an imaginary narrative consistent enough to maintain a stable concept of self. Social psychologists and anthropologists have also shown that the concept of self varies enormously from culture to culture and that selves are constructed in the domain of social interaction – 'there cannot be mirrors in the mind without mirrors in society'.

Recent publications, notably James Austin's *Zen and the Brain*, have tried to integrate philosophical scepticism about the notion of the self with insights from Eastern meditation practice and experimental findings from cognitive neuroscience. Austin's dictum – 'consciousness evolves when the self dissolves' – encouraged the American neuroscientist Andy Newberg to try to pinpoint the brain mechanisms associated with the sense of dissolving self traditionally associated with meditation practice and religious mysticism.

The third problem of consciousness, closely related to the issue of the self, is the problem of agency. If the self is a narrative fiction, then who is the author of conscious volitional acts? And, more importantly, how is it possible to reconcile our sense of free agency with the clear evidence that the agent must be subject to the (deterministic) laws of physics?

There is little agreement as to the degree of overlap between the three problems of consciousness. To Chalmers, sentience may even be a property of all functional systems (including thermostats), whereas selfhood and free action are restricted to sophisticated systems with language. By contrast, Searle argues that consciousness is a natural property of complex biological systems, and claims that the real explanatory gap is in the area of volition and free action. Dennett, meanwhile, sees the alleged conflict between determinism and free will as just another of those fake dichotomies that arise when we fail to take the implications of evolutionary design all the way to the wire.

Over in the chaos corner, Freeman and his acolytes claim that these are all pseudo-problems caused by a refusal to accept Hume's claim that the notion of 'causation' is a construct of human psychology.

Ten years on from the launch of the 'Decade of the Brain', perhaps we can claim that there has been a little conceptual clarification in the field of consciousness studies, but the final frontier of scientific understanding still looks as distant as ever.

Keith Sutherland
Executive editor of the Journal of Consciousness Studies

What is consciousness?

Susan Blackmore

*Visiting lecturer in psychology at the
University of the West of England*

Why am I here? Who am I anyway? Why does everything feel, and look, and hurt like this? I have been asking questions like this (or they have been asking me) ever since I can remember. For many years I thought I could find out by pursuing the paranormal – a fruitless task if ever there was one. Now the questions seem to converge on one big question: what is consciousness?

The problem of consciousness is real, and deep, and not quite like any other. I fell happily into it yesterday, walking high on the Devon cliffs, with the seagulls crying overhead. The grass brushing against my boots was so, well, grassy. It was green and lush and glistening, and changing all the time as I strode along. This grassiness was my experience. Only I had exactly that vision from exactly that point of view. Yet – and here is the problem – I also believe that there is real green grass growing on that cliff, that I have objectively real eyes that take in light, and objectively existing brain cells in my head that make me see. But how can this be? How can objective things like brain cells produce subjective experiences like the feeling that 'I' am striding through the grass?

This gap is what David Chalmers calls 'the hard problem'. Victorian thinkers called it the 'great chasm' or the 'fathomless abyss'. It is a modern version of the ancient mind/body problem – but it seems to get worse, not better, the more we learn about the brain. Neuroscience is rapidly explaining how brains discriminate colours, solve problems and organise actions – but the hard problem remains. The objective world out there, and the subjective experiences in here, seem to be totally different kinds of things.

Asking how one produces the other seems to be a non-sense.

This is what makes the problem of consciousness so interesting – and so painful. If you don't find it painful (and I will not apologise for wanting you to), pick up any object, such as a cup of tea or a pen, and just look. Do you believe there is a real cup there? Are you not also having a private subjective experience of the cup? How can this be? Call me a masochist but I like to induce this kind of pain in myself many times every day.

The intractability of this problem suggests to me that we are making a fundamental mistake in the way we think about consciousness – perhaps right at the very beginning. So where is the beginning? For William James, whose 1890 *Principles of Psychology* is deservedly a classic, the beginning is our undeniable experience of the 'stream of consciousness' – that unbroken, ever-changing flow of ideas, perceptions, feelings, and emotions that make up our lives. These thoughts and feelings flow by and 'I' experience them as they pass. This 'stream' seems to be what needs explaining.

But what if it is not like that? What if there is no stream? Can we even conceive of this possibility? Some recent experimental results suggest we might have to. These experiments reveal what is called 'change blindness'. Imagine you are looking at a complex scene – perhaps a street you can see from your window. You probably imagine that in your stream of consciousness is a rich and detailed representation of the trees and cars and people and buildings outside. Many times a second you move your eyes or blink but the picture seems to stay there. You probably imagine that if something changed you would notice the difference. You are probably wrong.

In change blindness experiments, a scene like this is shown to people but, by using clever eye trackers or other techniques, something in the picture is changed at the exact moment when they move their eyes. For example, a tree might disappear, a couple appear on the pavement, or a car be swapped for a van. In my own experiments, and many others, people typically fail to detect the change.

This is weird. If the change is made when their eyes are still, people notice the change immediately. This is because we have special detectors in the brain designed to notice objects that move, and draw our attention to them. But these detectors cannot work when the whole eye moves. If the change happens to an object the eyes are directly attending to, they notice it too, but otherwise it is as if nothing happened.

This peculiar effect cannot be dismissed as a quirk of lab conditions. Dan Simons, at Harvard University, showed the same effect in disturbingly ordinary situations. An experimenter approached a student on the campus and asked for directions. Meanwhile, two men picked up a door and carried it between the experimenter and the student as they talked. Hidden behind the door was a second experimenter who jumped up and took the place of the first. So now the poor student was talking to a completely different person. Amazingly, most of the time the students did not notice the substitution but just went on giving directions as before.

The conclusion seems to be this: we do not have in our heads a rich, stable and detailed visual image of the world at all. At any time we see in detail only the tiny area we are looking at. When we move our eyes, the detail is all thrown away, leaving at most a sketchy memory of the scene. We think it is all in our stream of consciousness because if ever we forget something we can just look again and there it is. We can use the outside world as a memory, so our brains do not need to keep the details. This way we get the illusion that the details are always there. This alone shows we are wrong about our stream of consciousness.

This has been called the Grand Illusion theory, but why should we suffer such an illusion? The answer may simply be that there is too much information out there for the brain to keep it all — think of how much computer memory a single picture takes up. But the illusion is deeper still.

Have you had this experience? The phone rings, or

the clock chimes, several times before you notice it. At that point you can distinctly count the number of chimes since it started – chimes you did not consciously hear. Or what about this? You drive a familiar route, and on arriving at your destination remember nothing of all those lights you stopped at, pedestrians you avoided and decisions you made. Obviously you were behaving extremely intelligently – otherwise you would be dead – but somehow 'you' were elsewhere – listening to the radio perhaps, or chatting with a passenger.

At any point in this journey, you might have suddenly woken up, as it were, and been sure that you had been perfectly conscious for the last few minutes. The odd occasions are when this does not happen and you realise how long the blank must have been. This suggests to me that we live our ordinary lives in a kind of daze. From time to time something wakes us up. In that moment of awakening, the brain concocts, from memory, a backwards story about what we were just experiencing. A stream of consciousness and a self who observes it both appear together – and both are illusions.

Illusion is the right word. An illusion is something that exists but is not what it seems. So the 'me' that seems to be steadily experiencing this world is not nothing, but nor is it the persistent observer with consciousness and free will that it seems to be.

How can I say that 'I' am an illusion? Surely I, Sue Blackmore, must have a self like you do? Well, yes and no. After decades of thinking about it, funny things happen to the sense of self. Not only have I struggled with the results of experiments like these, and practised living without free will, but I have also spent a lot of time sitting still and watching. The harder you look for the self who is experiencing things, the less obviously it exists. Indeed, there can arise states in which self and other are not separate at all. This is hard to describe, but is obvious when it happens.

I think we have a long way to go to see through these illusions but this is what we have to do. We need both to carry out careful experiments and to practise looking determinedly into the nature of experience itself. Perhaps then we will not see a stream of consciousness and a self who experiences it but how things really are. Only then will the hard problem disappear and the fathomless abyss close up.

Further reading

Susan Blackmore: *Beyond the Body* (Heinemann, 1982), *Dying to Live: Near-Death Experiences* (Prometheus Books, 1993), *In Search of the Light: The Adventures of a Parapsychologist* (Prometheus Books, 1996)

Ned J. Block, Guven Guzeldere and Owen Flanagan: *The Nature of Consciousness* (MIT Press, 1997)

Thomas Metzinger: *Conscious Experience* (Imprint Academic, 1996)

Steven Rose (ed.): *From Brains to Consciousness? Essays on the New Sciences of the Mind* (Penguin, 1999).

What is a Thought

When René Descartes declared, 'I think, therefore I am,' he put thinking on a pedestal where science could not reach it. He divided the universe so that every material thing could be measured and weighed and counted, but thoughts belonged beyond the quantifiable realm of time and space. Consequently, the seventeenth century – a scientific golden age that included development of anatomy and the discovery by physician William Harvey of the circulation of blood – saw no attempt to link the thinking process to human biology. That taboo lasted the best part of three hundred years.

In addition to philosophical prejudice, the study of mind/brain correlation was hampered by technical problems. How would one set about examining a thinking brain? A significant breakthrough came in the 1860s and 1870s, when post-mortem reports linked disruption of certain patients' language abilities with damage to certain areas on the left side of their brains. Ever since then the study of patients with selective brain damage, either from accident or illness, has remained an important source of evidence for brain scientists.

While physiologists examined the brains of dead subjects, Wilhelm Wundt (1832–1920) – the 'father' of experimental psychology – delved into the minds of living ones by the method of introspection. For instance, he took the speed with which volunteers reported seeing a flash or hearing a buzz as a measurement of the time it took for a sensation arriving at the eye or ear to register as a conscious thought. In another set of experiments, subjects were asked to detail the contents of their thoughts. Wundt supposed that there must exist a set of basic mental sensations that combined in various ways to create the whole range of thoughts and feelings. The aim was to produce an equivalent to chemistry's periodic table of elements, setting in order the array of mental 'elements' underlying the complex mental states that make up unified conscious experience.

The appearance of scientific rigour gave this project initial popularity, but disparities between subjective reports, and a public falling-out among Wundt's followers, brought introspectionism into disrepute. In a sharp reaction, psychology swung to the opposite extreme and embraced behaviourism, the doctrine that externally observable behaviour was the only truly scientific measure of what (if anything) was going on in people's minds. That remained the dominant doctrine in psychology for fifty years from the end of the First World War.

Fortunately, while psychology refused to study thought, things were at last progressing on the physiological front. The brain contains no pain-sensitive nerves, so – alarming as it sounds – it is possible to open up the skull under local anaesthetic and operate on the cortex (Hercule Poirot's 'little grey cells') while the patient is alert. From the early 1930s onwards, the Canadian brain surgeon Wilder Penfield exploited this fact to talk with his patients (who included his own sister) during surgery. Of particular interest was his treatment of epilepsy. Penfield would locate areas of cortex suspected of causing seizures by stimulating the likely sites with a mild electric shock. Each probe would cause the patient to have an unbidden thought, and when the reported thought matched one previously associated by that patient with the onset of a fit, Penfield knew he had found a problem area. As a by-product of this diagnostic work, he began to gather a large amount of evidence that showed certain areas of the brain's cortex were associated with specific kinds of thoughts or sensations.

Today's high-technology brain scanners provide a less invasive way for neuroscientists to observe what the brain is doing while volunteer subjects undertake various mental tasks. The computer images produced by these machines show different parts of the brain 'lighting up' and it seems almost as if we can now 'see people thinking'. But that is a mistake. The scanners give only indirect evidence since their readings reflect only the relative rate of blood flow at different points in the brain and the level of cortical activity has to be deduced from that. Even so, these techniques applied in experimental psychology, together with more direct work on animals and theoretical ideas being explored at the cutting edge of computer development, are contributing towards the first genuinely scientific study of thinking, called cognitive science.

The basic model of thinking used by most cognitive scientists treats the brain as an information processor. As a video camera picks up patterns of light waves and air pressure, combining and transforming this information by electronic means to produce a sound/vision tape recording, so – in a similar but more complex way – the human senses are assumed to pick up information from the outside world that is then combined and transformed by the brain into the mental states we call thoughts. Rather as a computer is programmed to respond in a certain way to particular inputs, so the brain's nerve cells (known as neurons) pass on information by reacting (or 'firing') to input signals from the sense organs.

As early as the 1940s, with the first suggestion that neurons acted like a computer circuit, it was realised how complex the brain's system must be. There are vast networks of cells, so that one neuron firing affects thousands of others. Feedback loops mean that when neuron A stimulates neuron B to fire, this might in turn stimulate A to fire again, causing an explosive effect that reinforces the original signal. Other signals, lacking reinforcement, will just peter out. The pattern of cells firing in response to different inputs provides the main focus of research in cognitive science. Are clusters of neurons 'hard-wired', so that they will always act in concert? Or is there 'plasticity', enabling thought and behaviour to vary even in response to identical inputs? Does learning perhaps consist of forming and maintaining certain networks? The answers to such questions hold the secrets of thought, of memory, of perception, of our whole mental life.

From being the research area that dare not speak its name, cognitive science has in the past decade become an unstoppable bandwagon. On the one hand it has attracted high-profile names from other fields, such as Nobel laureate Francis Crick of DNA fame. On the other, it has launched previously unknown neuroscientists, such as Susan Greenfield, into new careers as best-selling authors and media personalities – and, in her case, even into the House of Lords.

Despite its new-found popularity and apparent confidence, cognitive science is still long on questions and short on agreed answers. Take memory, for example. During the 1960s, a 'multi-store' model of memory was developed and popularised. All information was understood to come briefly into a sensory store, pass rapidly into a short-term store, and then, where appropriate, into long-term store. Within ten years, this scheme was shown to be grossly oversimplified. The short-term store was replaced with the concept of working memory, and long-term memory split into two distinct kinds: episodic and semantic. Yet all these neat distinctions prove to have exceptions and with new findings further problems abound. Another area concerns problems such as how someone can forget their name but remember how to frame the sentence, 'I have forgotten my name.' A proposed distinction between 'implicit' and 'explicit' memory sounds impressive, and has been widely accepted, but does not actually explain anything.

Meanwhile, the whole enterprise of treating the mind/brain like a computer that processes information still finds opposition at the highest level, from scholars as diverse as Oxford mathematician Roger Penrose and philosopher of mind John Searle.

Searle has a distinctive position, combining physicalism ('brains cause minds' is his slogan) with an insistence that mental states have an independent existence and cannot be 'reduced' to brain states. This sets him in opposition both to those who identify mental states (without remainder) with physical brain states and also to functionalists, such as philosopher Jerry Fodor, who say that mental states are embodied in the pattern or relationship of components in a physical system. For functionalists, the material of which a system is made is irrelevant to the mental states it supports. The components can be human brain cells, computer chips or old tin cans, so long as they are linked in the right way and information can pass appropriately between them. Both Searle and Penrose, using different arguments, claim that mere organisation of this kind ('syntax' in the jargon) can never impart the meaning or understanding that are essential to the conscious thinking process.

This leads into the really big question – David Chalmers's 'hard problem' – of how the universe gives rise to conscious thought at all. If consciousness evolved in time, then the capacity for it must have been present in the universe from the beginning. This has led some (including Chalmers himself) to admit some form of panpsychism, the belief that physical matter itself – right down to the level of sub-atomic particles – has the potential for consciousness. To avoid this conclusion, others – such as the husband-and-wife team of neurophilosophers Paul and Patricia Churchland – have jumped in the opposite direction and denied the reality of mental states altogether. A final – but little heard – possibility might be a return to Descartes and the placing of conscious thought in a different realm from the physical world.

Anthony Freeman
Managing editor of the Journal of Consciousness Studies

What is a thought?

Susan Greenfield
*Professor of pharmacology at Oxford University
and director of the Royal Institution*

This question appears at first to be impregnable. But perhaps this is merely because the particular phrasing has turned a familiar process – a verb – into an elusive thing – a noun. Given that we neuroscientists cannot root around in the brain looking for independent, monolithic structures that obligingly equate with mental processes, I prefer to translate 'a thought' into 'the process of thinking'.

One certainty is that this sophisticated and awesome mental process can be dissociated from raw consciousness, as in 'I did it without thinking'. Do people, for example, actually have thoughts while bungee-jumping? I for one, when engaged in the far gentler behaviour of dancing, fancy in retrospect that I was fleetingly a mere sponge to my senses, a passive recipient of music and beat, having a literally sensational time. Moreover, you only have to look at a baby suckling at the breast, or at a goldfish opening and closing its mouth, staring dully back at you from inside its bowl, to conclude that thought seems to have evaded fish and infant alike, even though they are presumably experiencing some kind of conscious state.

Then again, we should not be too hasty in disenfranchising non-human animals from any type of thinking and planning. Psychologists make their living out of the ingenuity and cognitive powers of laboratory rats and their extensive studies have shown that animals can devise strategies to achieve certain goals. Monkeys and chimps can plan ahead for hours – some researchers claim even for days. It is just that our primate cousins cannot plan for as long ahead as we can. Nor can they take thinking to its limit and

Big Questions in Science

contemplate their own demise. So we cannot say that thinking is something exclusive to humans and proceed from there.

Yet there are types of mental function, beyond global consciousness, that occur in both humans and other animals in differing measure and that we need to clear out of the way in our quest to pin down real thinking. For example, a mere association of pleasant or unpleasant experiences with a particular object, person or activity is a passive process that occurs unbidden and without effort. And, in another vein, the plodding, step-by-step, ruthless reasoning of the chess-playing computer – indeed of computers in general – hardly chimes with what most of us would be trying to capture when describing thought. As the atomic physicist Niels Bohr once admonished a student: 'You are not thinking, you are just being logical.' We all like to think that a thought is more than the end product of an algorithm, that it is something somehow more original. So how could this wonderful process occur in the brain?

When any neuroscientist starts to think about original, active thinking – 'cognition' – they will almost certainly point to the cortex. This is an area of the brain that wraps around the outer surface like bark around a tree – hence its name, after the Latin for bark. The cortex is the thin outer layer of the brain and is for many the repository of 'higher' functions, beyond brute subconscious processing of the senses and autopilot-type movement coordination. The reason for endowing the cortex with the most sophisticated functions is straightforward: the surface area of this region expands disproportionately during evolution. In the primate brain, it is heavily wrinkled to accommodate the surface area within the relatively tight confines of the skull, while in the brains of, say, the rat or rabbit, the texture is completely smooth. In the last trimester of pregnancy, the surface of a developing human cortex is transformed from completely smooth to resembling a walnut, due to the astonishing proliferation of cells at that time – up to 250,000 a minute.

Notwithstanding the fact that the cortex expands

with the sophistication of a species – and indeed that damage to the cortex can result in severe mental impairment – we still need to avoid the trap of regarding the cortex as the centre 'for' thinking. A cortex placed in a dish will not be very impressive intellectually; it is its contribution to the operations of whole brain organisation that is the critical issue.

So far, we scientists have been able to point to certain parts of the cortex that seem more involved in 'thinking' than others, and indeed have highlighted certain specific cognitive scenarios where parts and sub-parts of the cortex are highly active. But I would challenge any of my colleagues to describe what is actually happening amid the bump and grind of neurons that correlates with the process of active, original thought. The cortex, as part of the brain, is necessary without being sufficient and its micro-structure and intrinsic operations prompt little insight in pursuing our basic question any further.

Somehow the more ingenious the species, the more its brain will enable an animal to learn from and thus 'think' about experiences, rather than being at the mindless dictate of its genes. Even Craig Venter, the boss of the genomics company Celera, has admitted that the 30,000 genes in the body can in no way account for all our body and brain traits. This disparity is particularly striking in the brain, where counting the connections in the cortex alone, at one a second, would take some thirty-two million years. Moreover, connections in the cortex can reflect experiences, so the more individual experiences you have in life, the more you develop as an individual.

This 'plasticity' of the brain was demonstrated in a fascinating study of London taxi drivers, who have to possess 'The Knowledge', a memory for all the streets of London and how to navigate them. In this experiment, scans of cabbies' brains revealed that the architecture of a particular region of the brain relating to memory had changed, compared to the brains of non-taxi drivers of a similar age. Other studies have shown that practising the piano can change the brain territory relating to digits and,

Big Questions in Science

even more amazing, that mere mental practice can have a similar effect.

The brain has an astonishing ability not only to adapt to certain skills and activities, but also to compensate for damage after strokes and, on a more humdrum daily basis, to change behaviours and attitudes in the light of experience. We should therefore surely be looking not at anatomically, rigidly demarcated macro brain regions, nor at the micro level of inflexible genes that are simply switched on or off to manufacture a specific protein, but at an intermediate dynamic level: connections between networks of brain cells that band and disband. Perhaps it is here, within the restless and reactive circuits of brain cells, that we will find the most appropriate infrastructure for thought.

My own view is that these networks of neurons, which gradually develop and remain dynamic throughout the lifetime of the human brain, underscore for each of us our own individual take on the world. They imbue every process, person and object we encounter, with a particular, individual 'meaning'. It is this personalisation of the brain that in my view most happily fits our concept of 'mind'. In our everyday language, we selectively use the term 'mind', as opposed to 'brain', not because it is some intangible airy-fairy humanities indulgence contrasting with the squalor of biomedical science, but because phrases like 'broad-minded', 'change your mind' and 'I don't mind' all emphasise our own individual perspective.

Now consider 'losing my mind', 'blowing your mind', and 'letting yourself go'. In such scenarios, thinking is minimal as we are reduced to being passive sponges to our senses, recapitulating the 'booming, buzzing confusion' of the small infant or perhaps even of the goldfish. These states of visceral, raw feeling, of 'pure' emotion and zero thinking are, as I see it, the building blocks of consciousness and can therefore occur in the absence of the thought process.

However, most of the time we are not in mid-orgasm, nor experiencing road rage, nor gasping in a bungee jump, nor wallowing in the sensuality of a fine claret, nor being driven by the throbbing beat of music. Instead, we are evaluating the world, and accordingly inter- acting with it. In such cases, our covert and idiosyncratic connections will be at work, colouring our attitudes to whatever we encounter, forming the basis for the 'sub- conscious' mind.

But we are still hankering after the creative and orig- inal process of thinking. 'I have just had a thought' is syn- onymous, after all, with having an idea. In this case, my suggestion, and indeed answer to the question, is that hubs of connections that were previously independent hook up with each other, as networks of neurons can. This coalescence results in a new and perhaps unusual juxta- position of association. One extreme example would be the thought of US theoretical chemist and biologist Linus Pauling, who imported the principles of physics into chem- istry, with his masterly concept of the chemical bond, or Australian immunologist Macfarlane Burnet, who saw a similarity between the principles of Darwinian evolution and the immune system. It is this breathtaking leap between one scenario and another, between one discipline and another, that surely is the true apotheosis of the human mind, and is A Thought at its finest.

Further reading

Susan Greenfield: *The Human Brain* (Phoenix Press, 1998), *Brain Story* (BBC Consumer Publishing, 2000), *The Private Life of the Brain: Emotions, Consciousness and the Secret of the Self* (John Wiley & Sons, 2001)
William Lyons: *Matters of the Mind* (Edinburgh University Press, 2001)
John McCrone: *Going Inside* (Faber & Faber, 2000).

What is a Dream?

? Dreams fascinate us. Whether they are recurring nightmares or benign, surreal images, we are desperate to know what they mean. If only we could interpret them, we believe, we could unlock the hidden depths of our personalities and show the world what interesting people we truly are.

They have also fascinated different ages and cultures, playing a huge role in literature, philosophy and religion. The earliest recorded dreams date back to the Sumerians of Mesopotamia in 3100 BC. Aristotle thought dreams, being absent from external stimuli, could heighten awareness of internal sensations. Artemidorus, author of the first *Interpretation of Dreams* in the second century, classified dreams into two types: prophetical and realist, that is, dreams which deal with current concerns and are affected by the dreamer's state of mind and body. From the Sumerians and ancient Greeks, through to modern psychotherapy, people have seen dreams as a heightened form of communication – as divine messages, creative inspiration, prophecies, or keys to hidden desires. Yet many have also dismissed them as nonsense. Francis Bacon wrote in his *Essays* in 1625: 'Dreams and predictions of astrology . . . ought to serve but for winter talk by the fireside.'

It is only relatively recently that dreams have moved from the realms of the mystical and literary into the world of science, and a glance at the numerous dream exchange sites on the Internet shows that the mystical still holds sway. Sigmund Freud was tackling the science of dreams around 1900. In his *Interpretation of Dreams*, he described dreams as 'the royal road to the unconscious', an unconscious made up of (mainly sexual) desires that had been blocked out or repressed by waking thought. By getting patients to talk about their dreams, Freud uncovered the tension between conscious and unconscious thought, and linked the resulting repression to mental illness. His book was the first systematically scientific work on how the mind works.

At first, Freud used hypnotic suggestion to get at the unconscious, trying to induce a dreamlike state in his patients. He then moved on to what later became known as 'free association', simply asking his patients to say the first thing that came into their head and to connect ideas loosely from there. This provided insights into their unconscious mind, although he came up against resistance from the patient's conscious will. From this, Freud developed his theory of the three-dimensional mind, divided between the id, which applied to instinctive thought; the ego, which applied to organised, realistic thought; and the superego, which covered moralising and critical functions.

Freud's work had a huge impact on ideas about identity, creativity and mental health, although some of his theories may now appear dated or subjective. In literature, for example, his work was crucial to the surrealist movement, whose followers used his ideas about the unconscious in their own experimentations on writing. Many writers practised what became known as automatic writing, which was supposed to come directly from the unconscious and represent unblocked desire.

However, Freud was not the only scientist to show an interest in dreams at the turn of the twentieth century. The Dutch psychiatrist and novelist Frederik van Eeden also developed a theory of the dream that still has an impact today, although many of his conclusions have been contradicted. He is credited with coining the term 'lucid dreaming', meaning dreaming while knowing that you are dreaming. Van Eeden began studying his dreams from 1896 and wrote down the most interesting in a diary. In 1898, he began to record a particular type of dream, later dubbed the lucid dream. In 1913, he published *A Study of Dreams*, in which he distinguishes nine different types of dream, from initial and pathological dreams to lucid and symbolic dreams. He believed that dreams were not purely arbitrary, and that there must be some kind of scientific order behind them. 'To deny may be just as dangerous and misleading as to accept,' he stated.

Freud's greatest contemporary and rival in terms of dream theory was Carl Jung. At first an ardent follower of Freud, Jung is known to have had several lucid dreams and occasional visions, such as his 1913 vision of 'a monstrous flood' engulfing Europe and leading to the destruction of civilisation. Like van Eeden, he carefully recorded his dreams and visions and described various archetypal figures that appeared in them. According to Jung's theory, the psyche is made up of three parts: the ego or conscious mind; the personal unconscious, which includes memories that spring easily to mind as well as those that have been repressed for various reasons; and the collective unconscious – a kind of reservoir of human experience that influences all human behaviour, particularly emotional behaviour, and includes recognition of some key symbols and myths. As with

Freud, the dream was a tool in uncovering the hidden realms of the psyche and resolving emotional and other problems.

Later scientists, such as Fritz Perls, founder of gestalt therapy, developed Jung and Freud's work on dreams along phenomenological, subjective lines. Like Jung, Perls was interested in the emotions, but, rather than linking dreams to repressed memories, he focused on his patients' immediate situation. His aim was to restore a sense of wholeness to people he believed had been separated from uncomfortable thoughts, sensations and emotions. He got patients to act out their dreams and to comment on their emotions as they were doing so. The ensuing dialogue between dreamer and the dream object they were acting out aimed to break down the barriers holding the patient back from self-discovery.

Another key twentieth-century figure was psychologist Calvin Hall, who, from the 1940s until he died in 1985, collected over 50,000 dream reports. In his work, which prefigured the later trend towards a cognitive approach to psychology, he looked for patterns in dream content and developed a quantitative coding system by which he divided dreams according to settings, characters, emotions, objects and so forth. According to his theory, dreams express 'conceptions' of the self, family, friends and social environment, and the frequency of certain elements of dreams reflects waking concerns. His research on people around the world showed remarkable similarities between dreams, and long-term follow-ups of patients revealed great consistency in their dream content, with any changes reflecting alterations in their waking life. This continuity between dream content and waking thought was so marked that Hall felt he could predict a dreamer's behaviour and lifestyle through analysing the contents of their dreams.

The 1960s saw a reawakening of interest in dream theory, with work by scientists such as the psychotherapist Eugene Gendlin and the Jungian psychologist Arnold Mindell, whose Dreambody theory integrates dreams, the body and meditation. Mindell argues that 'all body problems, all body symptoms are dreams trying to manifest through the body' and that 'accurately following the way of nature' and amplifying the signals from dreams brings awareness of the patterns on which we structure our lives, including those which are painful and which we may have repressed.

In recent years, the focus of research has oriented more

towards the physiological. Sleep laboratories have been developed, where patients can be viewed in controlled conditions and tests performed to show how the body undergoes the same chemical reactions to dream images as to waking reality. Another development has been brain-imaging technology, which shows that the brain is just as active when asleep as when awake.

In the 1950s, the University of Chicago conducted a series of studies on rapid eye movement (REM) sleep that showed a link between dreams and the various stages of sleep. They found that sleepers could recall dreams most frequently if they were awakened when their eyes appeared to be moving rapidly beneath their eyelids.

Since then, although there has been much work on the content of dreams – including the effects of gender and emotions on dreaming – the emphasis has shifted more on to the dreaming process. As computers became more popular, it was suggested by several scientists that REM served the same function as the scan disk on a computer, preparing the brain for the day ahead or processing the events of the day before into the memory database. Biologists Francis Crick and Graeme Mitchison caused a stir in 1983 when they argued in the journal *Nature* that 'we dream to forget'. Their view was that dreams are the unnecessary and harmful detritus of the overload from everyday life, and REM sleep is a kind of binman of the mind.

Recently, David Maurice, professor of ocular physiology in the department of opthalmology at Columbia Presbyterian Medical Center, has published research suggesting that the function of REM sleep is to ensure a supply of oxygen to the cornea of the eye during sleep, rather than helping to process memories of the preceding day's events. Others suggest a genetic role. Computer programmer Bradley York Bartholomew, a member of the Association for the Study of Dreams, believes REM triggers the action of a gene that may programme the neurons of the brain for specific functions in waking consciousness.

There has also been growing interest in non-REM dreaming. Ernest Hartmann, a former president of the Association for the Study of Dreams, stresses that, while investigations around REM sleep mean that we now understand the biology of the underlying state which offers the best conditions for dreaming, we still do not know what dreaming is. He believes that dreams function in an auto-associative way in the outer reaches of the brain, guided by emotion, whereas conscious thought works on a more limited area involved in general input and output of information. His argument, based on his work on trauma patients, is that the role of dreams is to weave new experience into the memory and increase the connections between events. This would fit in with work on infants that suggests that nightmares set in at the toddler stage when children have amassed enough of a memory bank to create connections.

Another development has been the central position accorded to dreams in the emerging science of consciousness. For example, research on conscious dreaming techniques, such as that carried out by Stephen LaBerge's Lucidity Institute, is giving new insights into the workings of the mind. The institute has shown through controlled clinical studies that people have the ability to stay conscious during sleep. The aim is to use this ability for personal development, thereby combining some of the elements of classical dream theory with the scientific lab-based experimentation of the biological approach.

However, we are still a long way from understanding the mysteries of the mind. As Patricia Garfield, another ASD former president, says: 'Although we have learned a lot about the mechanics of dreams this century – the pattern of dreaming during the night and its physiological components – we have made little progress in understanding the meaning of our dreams.' The fascination continues.

Mandy Garner
Features editor of the Times Higher Education Supplement

What is a dream?

Stephen LaBerge
Director of the Lucidity Institute, Palo Alto, California

Every night we enter another world, the world of dreams. While we are dreaming we usually implicitly believe that we are awake. The mental worlds of dreams are so convincingly real that we mistake them for the 'outside world' that we share with other people. How can this be? Why does it happen? What is the relation between our day and night lives? And what is the origin and function of dreaming? Incredibly, in this age of scientific understanding of the most intricate workings of biology, there is little scientific consensus about the answers to these questions.

The *Oxford English Dictionary* defines a dream as 'a train of thoughts, images, or fancies passing through the mind during sleep'. But such a definition fails to capture the lived-in, experiential reality of dreams. In my view, dreams are much more accurately described as experiences – that is, conscious events that one has personally encountered. It may seem odd to speak of dreams as conscious experiences, but the essential criterion for consciousness is reportability, and the fact that we can sometimes remember our dreams shows them to be conscious rather than unconscious mental processes. We live through our dreams as much as our waking lives. In these terms, dreaming is a particular organisation of consciousness.

Of course, that begs the question what is consciousness? For me, it is the dream of what happens. Whether awake or asleep, your consciousness functions as a simplified model of yourself and your world constructed by your brain from the best available sources of information. During waking, the model is derived from external sensory input, which provides the most current information about present circumstances, in combination with internal contextual, historical and motivational information. During sleep, little external input is available, and given a sufficiently functional brain, the model is constructed from internal biases. These will be expectations derived from past experience and motivations – wishes, for example, as Freud observed, but also fears. The resulting experiences are what we call dreams, the content of which is largely determined by what we fear, hope for and expect. From this perspective, dreaming can be viewed as the special case of perception without the constraints of external sensory input. Conversely, perception can be viewed as the special case of dreaming constrained by sensory input.

There are two kinds of sleep: an energy-conserving state known as Quiet Sleep (QS), associated with growth, repair and restoration, a relaxed body and a brain on idle, and a very different state known variously as Active Sleep, REM or Paradoxical Sleep (PS). This is associated with rapid eye movements and muscular twitches, a paralysed body, a highly activated brain and dreaming. Although PS is not the only sleep state in which people can dream, it provides the optimal conditions for vivid dreaming – a switched-on brain in a switched-off body.

It is frequently assumed that waking and dreaming experiences are completely distinct. Dreams, for example, are said to be characterised by lack of reflection, lack of control over attention and the inability to act deliberately. But the evidence flatly contradicts this characterisation of dreams as single-minded and non-reflective. In recent studies directly comparing reports from waking and dreaming, my colleagues Tracy Kahan, Lynne Levitan and Phil Zimbardo and I found that, compared to waking experiences, dreaming contained public self-consciousness and emotion slightly more frequently, and deliberate choice slightly less frequently. However, no significant differences between dreaming and waking were found for other cognitive activities, and none of the measured cognitive functions was typically absent or rare in dreams. In particular, nearly identical levels of reflection were reported in both states.

The fact that dreams contain sudden shifts of characters and scenes of which the dreamer takes little note is sometimes cited as evidence for a cognitive deficiency in dreaming. The presumption is that if this occurred in waking, one would immediately notice and attempt to understand the discontinuity. However, this assumption is unwarranted. Recent studies on 'change blindness' have shown that people are far less likely to detect environmental changes than common sense assumes.

I am not saying that there are no differences between dreams and waking experiences. For example, the dream world is much less stable than the waking world because the dream lacks the stabilisation of an external structure – physical reality. Likewise, one can violate the laws of physics and society in dreams without the usual consequences. But the absence of sensory constraint is the only essential difference. One might or might not know that one is dreaming and it would still be a dream. And whatever differences there may be, I believe they are more alike than they are different. As the early twentieth-century English physician and writer, Havelock Ellis, said: 'Dreams are real while they last. Can we say more of life?'

It is likely, however, that PS evolved for purposes more basic than dreaming. Just as philosophy, poetry, music, or abstract mathematics are probably the lucky side effects of other features that emerged through natural selection – such as general-purpose language – so dreaming is perhaps something that humans do, and extract value from, but that did not evolve directly.

The distribution of PS across development and in the course of a night provides a clue to the most important functions of this sleeping state. PS is at its maximal level perinatally and in the last weeks of prenatal development when the brain is growing its vast networks of neural circuitry. The appealing idea that PS serves as an endogenous state for the unfolding of genetic programming has been proposed by French sleep researcher Michel Jouvet and Stanford University psychiatrist William Dement, among others. The percentage of PS gradually drops off throughout childhood, but does not completely disappear when brain growth stops at adulthood, implying that PS may serve another function. The fact that PS gradually increases across the night, reaching a maximum as the time of waking approaches, suggests that it may prepare our brains for waking action; a sort of brain tune-up. These recurrent activations every ninety minutes or so throughout the night may also help consolidate new learning.

One of the most characteristic features of dreams is how difficult they are to remember. The average person dreams at least six times a night, but recalls them only weekly. The explanation of why dreams typically are rapidly forgotten lies, once again, with evolution. Humans learn that dreams are distinct from other experiences by talking to other humans. But non-speaking animals have no way to tell each other how to distinguish dreams from reality. For them, explicit dream recall could cause potentially fatal confusion. The purpose of PS cannot therefore have anything to do with the explicit recall of dreams, to say nothing of dream interpretation. Nevertheless, since humans can tell the difference between dreams and waking reality, remembering dreams should do us no harm, and indeed may inspire us to remake reality in accord with our dreams.

Even if dreaming has no special biological function, dreams themselves may play a specific role. They may, for example, increase variability in the nervous system. Darwinian evolution requires a variable population, a selective pressure and a means of reproduction of successful variations. Perhaps dreaming generates a wide range of behavioural schemas or scripts guiding perception and action from which to select adaptations fitted to changing environments. Be that as it may, answers to why we dream need not be framed so narrowly. For some, the answer is: we dream to find out why we dream. Personally, I prefer: I dream to find out who I am beyond who I dream I am.

The view of dreams as world models is far from the

traditional notion of dreams as messages, whether from the gods or from the unconscious mind. Nonetheless, interpretation of dreams can be very revealing of personality, and a rewarding practice. If what people see in ink blots can tell something about their personal concerns and personality, how much more revealing should dreams be, because they are the worlds we have created from the contents of our minds. Dreams may not be messages, but they are our own most intimately personal creations. As such, they are unmistakably coloured by who and what we are and could become.

Our dreams seem so real that it is usually only when we wake up that we recognise them as the mental experiences they are. Although this is how we generally experience dreams, there is a significant exception: sometimes while dreaming, we consciously notice that we are dreaming. This clear-sighted state of consciousness is referred to as lucid dreaming.

During lucid dreams one can reason clearly, remember the conditions of waking life, and act voluntarily within the dream upon reflection or in accordance with plans decided upon before sleep – all while remaining soundly asleep, vividly experiencing a dreamworld that can appear astonishingly real.

Until recently, researchers doubted that the dreaming brain was capable of such a high degree of mental functioning and consciousness. In the late 1970s, our laboratory research at Stanford University proved that lucid dreams did in fact occur during unambiguous sleep. Based on earlier studies showing that some of the eye movements of PS corresponded to the reported direction of the dreamer's gaze, we asked lucid dreamers to carry out distinctive patterns of voluntary eye movements when they realised they were dreaming. The prearranged eye-movement signals appeared on the polygraph records during uninterrupted PS, proving that the subjects had indeed been lucid during sleep.

Subsequently, my colleagues and I began new studies of the dreaming mind, made possible by the ability of lucid dreamers to carry out experiments within dreams. We learned that the physiological effects on the brain and body of dream activities are nearly identical to the effects of experience in waking life. For example, we found that time intervals estimated in lucid dreams closely match actual clock time, that dreamed breathing corresponds to actual respiration, that dreamed movements result in corresponding patterns of muscle twitching, and that dream sex shows physiological responses similar to actual sexual activity.

In addition to providing an effective way to carry out scientific explorations of dreaming consciousness and mind/body relationships, lucid dreaming also offers considerable potential for a variety of other applications. These include aiding self-development, enhancing self-confidence, overcoming nightmares, perhaps improving mental health, facilitating creative problem solving and (as shown by the thousand-year-old practice of Tibetan dream yoga) opening the mind to the possibilities of higher development. The broadest appeal of lucid dreaming is likely to be the fantasy-come-true of being able to have any imaginable experience.

What currently limits the number of lucid dreamers to a growing minority is that while lucid dreaming is a learnable skill, it takes time and effort. Hence a major line of our research has been developing techniques and technology to make lucid dreaming more easily accessible. The 'Model-A' of this emerging 'oneirotechnology' is the NovaDreamer®, a biofeedback device that gives cues during PS reminding users that they are dreaming. New, increasingly effective lucidity induction devices are under development, bringing closer the dream of a personal world simulation for everyone.

Dreams have long been regarded as a wellspring of inspiration in nearly every field of human endeavour, from literature to science, engineering, painting, music and sports. August Kekulé's dream of a snake biting its own tail, to take just one example, inspired his discovery of the previously unsuspected ring structure of benzene. In the past, we have had little or no control over the occurrence of creative dreams. But it now seems possible that the fantastic and heretofore unruly creativity of the dream state might be brought within our conscious control by means of lucid dreaming. As Kekulé urged his colleagues on the occasion of presenting his dream inspiration to a scientific convention in 1890: 'Let us learn to dream . . .'

Further reading

Lucidity Institute http://www.lucidity.com

Michael Jouvet: *The Paradox of Sleep: The Story of Dreaming* (MIT Press, 1999)

Stephen LaBerge (with H. Rheingold): *Exploring the World of Lucid Dreaming* (Ballantine, 1990)

Anthony Stevens: *Private Myths: Dreams and Dreaming* (Hamish Hamilton, 1996)

R.L. Van de Castle: *Our Dreaming Mind* (Ballantine, 1994).

What is Intelligence?

For a quality traditionally associated with the mind, intelligence sparks strong emotions. What it is, how we identify it, how far it is inherited and how far acquired are some of the most controversial questions facing scientists today. They are questions that have implications not only for how we see ourselves but for how we bring up our children. Parents lap up advice about altering diet or playing Mozart to a developing foetus in an effort to secure for their offspring a characteristic now seen as more of a key to future success than beauty or physical prowess.

The role of nurture in intelligence has been debated since Plato, who argued that learning was of little use to those of low mental ability. But the issue came to the fore after Charles Darwin published his *On the Origin of Species*. In 1869, Francis Galton, a scientist and cousin of Darwin's, published *Hereditary Genius*, in which he analysed the histories of famous families and found there was a high probability of eminent people having eminent relatives. From this he concluded that intellectual ability was inherited – rather underestimating the fact that intelligent parents were more likely to provide a stimulating environment for their children. He believed that selective breeding would improve human cognitive powers.

Galton measured intelligence by testing reaction times to simple cognitive puzzles. Today's intelligence testing, which measures specific mental functions, originated in the early part of the twentieth century with French psychologist Alfred Binet. Schooling had just been made compulsory in France and Binet was asked to develop ways of identifying children likely to fall behind without special educational help. He drew up a series of mental tasks associated with everyday life, such as counting coins or memorising lists, designed to measure mental functions including memory, verbal ability and creativity. Children were judged by an overall score achieved across all tests. Binet also developed the concept of mental age – the age at which a normal child should be able to accomplish a particular task.

He insisted that his test was not designed to measure adults or to differentiate between children of normal intellectual ability. It was supposed to be an indicator of future school success only. But Binet helped to fuel the two great controversies in the study of intelligence. First, he established the disputed idea of a single, measurable characteristic of 'general intelligence'. Second, he plunged directly into the debate over intelligence's heritability – the extent to which genes are responsible for differences between individuals.

His work was quickly appropriated. Lewis Terman, an American cognitive psychologist at Stanford University, developed what became known as the Stanford–Binet test. This was more an examination of 'higher functioning' than a way of identifying backward students. It assigned to individuals a numerical intelligence quotient, or IQ, by dividing a person's mental age, identified by the test, by their chronological age and then multiplying by 100.

The notion of a single, general and measurable type of intelligence is usually referred to as 'g' and was first identified by psychologist Charles Spearman at the beginning of the twentieth century. Its supporters point out that IQ is a useful predictor of school success, is relatively consistent throughout life and that people who score highly in a test of one cognitive ability usually score highly in tests of another such ability. In the 1980s, Hans Eysenck, a German-born London University psychologist, claimed further evidence for g by showing a high correlation between IQ and reaction time. For example, people with high IQs asked to hit a button as soon as they saw a light go on, tended to hit it faster than those with lower IQs. Such a simple test appeared to eliminate the cultural, environmental and educational influences that critics of g say bring other tests of intelligence into question.

Those who believe in g believe it is something substantially heritable. Galton was the first to use twin studies to measure how much differences between individuals' intelligence depended on the environment and how far on genetics and to conclude that genetic factors dominated. Since then, many further tests have been carried out on identical and non-identical twins and siblings brought up in different environments, with varying conclusions. More recently, some researchers, Robert Plomin among them, have been working to identify some of the specific genes responsible for the heritability of cognitive abilities and disabilities. Although Plomin stresses that this kind of work involves looking at many genes rather than just one, the idea has disturbed social scientists concerned at the ethical implications.

But by no means all scientists believe intelligence is innate; estimates of IQ heritability still range from below 40 per cent to 80 per cent. British biologist Steven Rose says methods of calculating heritability are dubious because genetic and environmental factors are so difficult to separate. He argues that recent evidence showing rising overall IQ scores and declining differences in scores between populations suggests environmental factors must be crucial. Leon

Kamin, an American psychologist, is a similarly outspoken critic of IQ tests as a useful measure of intelligence and of the idea that intelligence is in the genes. He was the first to suggest that British educational psychologist Cyril Burt had fabricated data from twin studies to prove his theory that around 80 per cent of intelligence was inherited.

Fraud is not the only unpleasant allegation to enter the intelligence debate. Charges of racism have dominated since soon after Binet created his tests. Experiments in the early twentieth century by H.H. Goddard, director of research at an American school for the 'feeble-minded', showed that four-fifths of Hungarians, Italian and Russian immigrants arriving at New York's Ellis Island were below average intelligence. He advocated preventing 'morons', as he termed them, from bearing children. Tests carried out by Robert Yerkes for entry to the United States Army during the First World War showed similarly low immigrant scores, with black Americans coming bottom, and contributed to tough new immigration laws.

In 1969, Arthur Jensen published a long article in the *Harvard Educational Review*, which argued that genetic as well as social and environmental factors explained the fact that black Americans regularly score 15 per cent lower than whites in IQ tests. His views, which prompted an outcry, were taken up by psychologist Richard Hernstein and social scientist Charles Murray in *The Bell Curve: Intelligence and Class Structure in American Life*, published in 1994. This argued that economic and social power was determined by g, that most social problems were rooted in low intelligence, and that genetic differences between blacks and whites accounted for IQ differences and therefore for the way blacks were over-represented in the underclass.

The racism debate rumbles on, those opposing the *Bell Curve* view insisting that racial differences are predominantly due to cultural bias in the tests and differing socialisation. Curiously, there has been no similar debate over gender differences, in spite of apparent structural differences between male and female brains. IQ tests are controlled for gender, and men and women perform broadly the same – although there are more men at the top and bottom of the scales and men tend to perform better in spatial questions, while women perform better in verbal tasks.

These differences in type of task become more significant for critics of the g concept, who argue that there is not just one, general intelligence but several. One of the first to put forward this view was

Big Questions in Science

Louis Thurstone, an American electrical engineer and psychologist in the middle of the twentieth century. He identified a number of primary mental abilities an individual needed to survive and succeed. More recently, Howard Gardner, a professor of education and neuro-psychologist at Harvard University, has suggested seven types of intelligence. By studying individuals with disabilities, he localised parts of the brain needed to perform particular tasks and found seven different areas of the brain relating to, for example, musical, mathematical and linguistic functions. American psychologist Robert Sternberg has narrowed these multiple intelligences down to just three: analytic, creative and practical. He argues that psychometric tests can only measure the first of these.

In the past few years, researchers in the field have been studying yet another kind of intelligence – the artificial kind. British cybernetics professor Kevin Warwick has suggested: 'It is difficult to think of any area of human intelligence (even including that intelligence solely associated with being human) in which a machine will not soon be able to out-perform us.' Cambridge physicist Stephen Hawking has also warned that machines could one day develop intelligence and take over the world. He suggests that humans should be genetically engineered to improve their own intelligence and so prevent this happening.

But machines are a long way from acquiring one type of intelligence at least. In *Emotional Intelligence*, published in 1995, journalist and former academic Daniel Goleman suggested that a high IQ did not necessarily equal worldly success. In fact, individuals with very high IQs were often held back by emotional weaknesses. More useful as a life tool, he argued, was emotional intelligence, which he defined as a mixture of motivation, empathy and the ability to manage emotions and relationships. His ideas, while hugely popular with the general public and businessmen, tend to be dismissed by academics as unscientific. Which of these judges is the more valid depends, perhaps, on your definition of intelligence.

Harriet Swain
Deputy features editor of the Times Higher Education Supplement

What is intelligence?

Robert Plomin

*Medical Research Council research professor, Social, Genetic
and Developmental Psychiatry Research Centre,
Institute of Psychiatry, King's College, London*

The word intelligence means so many different things
that it is best to use other terms in order to avoid confusion. What I mean by intelligence is 'general cognitive ability', or g, which refers to the substantial overlap that
exists between different cognitive processes. This overlap is
one of the most consistent findings of research into individual differences in human cognitive abilities over the past
century. It has been found even in tests of processes that
seem to have little in common.

For example, general reasoning is assessed through
tests such as Raven's Progressive Matrices, in which subjects need to detect logical progression in a series of
matrices consisting of geometric forms. Spatial ability is
assessed through solving mazes, identifying simple geometric figures embedded in more complex shapes and
deciding whether or not one figure is a rotated version of
another. Vocabulary tests assess the product of previous
learning, while tests of memory typically involve presenting
digits or pictures to see how well they are recalled.

Despite this diversity, individuals who do well in one
test, tend to do well in others. In a meta-analysis by John
Carroll in 1993 of 322 studies that included hundreds of
different kinds of cognitive tests, the average correlation
was about 0.30, which is highly significant. This overlap
emerges not only for traditional measures of reasoning and
of spatial, verbal and memory abilities, such as those
mentioned above, but also for rates of learning and for
information-processing tasks that rely on reaction time.

Psychologist Charles Spearman recognised the
overlap in cognitive abilities nearly a century ago. He called

it g in order to create a neutral signifier of general cognitive ability that avoided the many connotations of the word intelligence. This g is best assessed by a statistical technique called principal components analysis. This identifies a composite dimension that represents what diverse cognitive measures have in common.

Such analysis indicates that g accounts for about 40 per cent of the total variance in people's performance of cognitive tests. The rest of the variance is accounted for by factors such as spatial, verbal and memory abilities, and by variance unique to each test. The more complex the test the higher the importance appears to be of the g factor. For example, g accounts for a high amount of variance in Raven's Progressive Matrices, while it is less important in simple memory, reaction or processing speed tests.

But g is not just a statistical abstraction. A look at a matrix of correlations between cognitive measures will show all overlap strongly while g is also indexed reasonably well by a simple total score on a diverse set of cognitive measures, as is done in IQ tests. In fact, g is one of the most reliable and valid traits in the behavioural domain; its long-term stability after childhood is greater than for any other trait, it is better at predicting important social outcomes such as educational and occupational levels, and it is a key factor in cognitive ageing. There are, of course, many other important non-cognitive abilities, such as athletic ability, and g by no means guarantees success either in school or in the workplace. Achievement also requires personality, motivation and social skills, now referred to as 'emotional intelligence'. But nothing seems to be gained by lumping all such abilities together as the popular notion of 'multiple intelligences' does. For me, g is what intelligence is about.

Although the evidence for g in the human species is widely accepted, acceptance is not universal. Arguments against it involve ideological issues, such as concerns that it merely reflects knowledge and skills that happen to be valued by the dominant culture, and objections of a more scientific nature. These include theories that focus on specific abilities, such as Howard Gardner's theory of multiple intelligences, which argues for very specific abilities, including non-cognitive abilities such as dance, and Robert Sternberg's 'componential' theory of cognitive processing, which attempts to identify cognitive processes that underlie cognitive abilities. When these theories are examined empirically, however, g shines through. For example, Sternberg concludes: 'We interpret the preponderance of evidence as overwhelmingly supporting the existence of some kind of general factor in human intelligence. Indeed, we are unable to find any convincing evidence at all that militates against this view.'

Concepts such as working memory are only just beginning to be assessed through traditional measures of cognitive ability, but so far they show the same. Of course, g is not the whole story – factors representing specific abilities are also important – but trying to tell the story of cognitive abilities without g loses the plot entirely.

The existence of g appears to go against the tide of current cognitive neuroscience that considers cognitive processes as specific and independent. But research in cognitive neuroscience focuses on average performance – for example, what bit of the brain lights up with neuroimaging when a particular task is performed – and g is not about average performance; it is about the individual differences in performance and the fact that individuals who perform well on some tasks tend to perform well on most tasks. In this kind of analysis, the data clearly point to g.

But the fact that g exists does not imply that its source must be a single general physical process, such as the complexity of a neuron's dendrites or the extent of the myelin sheath surrounding neurons' axons. Nor is its source solely physiological, involving synaptic plasticity, or speed of nerve conduction. Nor is it a single psychological process, such as working memory. Instead, it represents a chain of such physical, physiological and psychological processes, all enlisted together to solve functional problems. As an analogy, athletic ability

Big Questions in Science

depends on psychological processes such as motivation, physiological processes such as oxygen transport, and physical processes such as bone structure. But athletic ability is not one of these things, it is all of these things.

Genetic research is also important for the story of g since g is substantially heritable. There are more studies addressing the genetics of g than there are for any other human characteristic. Studies including more than 8,000 parent-offspring pairs, 25,000 pairs of siblings, 10,000 twin pairs and hundreds of adoptive families all come to the conclusion that genetic factors contribute substantially to g. Estimates of heritability vary from 40 to 80 per cent in each study, but estimates based on the entire body of data are about 50 per cent, indicating that genetic variation accounts for about half of the variance in g. Now research has begun to look for the specific genes responsible for this heritability – research that is likely to be accelerated now that the draft of the human genome sequence has been published.

But genetic research has gone beyond merely demonstrating that g is substantially heritable. A particularly important finding from multivariate genetic research, which analyses genetic and environmental sources of covariance among traits, is that g is where the genetic action is. While g accounts for about 40 per cent of the total variance of cognitive tests, multivariate genetic research indicates that g accounts for nearly all of the genetic variance of cognitive tests. That is, what is held in common among cognitive abilities is almost completely genetic in origin. What makes us good at all tests is largely genetic but what makes us better at some tests than others is largely environmental. This suggests that genetic links among cognitive processes may have been forged by evolution to coordinate effective problem solving across the modules of mind.

Genetic research also shows a strong genetic overlap between school achievement, as assessed by tests or grades, and g. In other words, the same genetic factors that contribute to individual differences in g are responsible for many individual differences in school achievement. Conversely, what is different between achievement and ability or g is largely environmental. These findings suggest that assessing achievement 'corrected' for g could bypass genetic influence, with far-reaching implications for educational selection, assessment and issues of added value in schools.

Another surprising finding is that the strength of genetic influence on g increases from early childhood, through middle childhood, to adolescence. This is counterintuitive to the notion that environmental influences accumulate as life goes by. It suggests to me that children actively select, modify and even create environments conducive to the development of their genetic proclivities. For this reason, I think about g as an appetite rather than an aptitude. At the simplest level this means that children like to learn what they find easy to learn. But I think that there is more to it than that: the mechanisms by which genes affect the learning process may have as much to do with motivation as with the brain's hard-wiring.

Further reading

Howard Gardner: *Frames of Mind: The Theory of Multiple Intelligences* (Basic Books, 1993)
Daniel Goleman: *Emotional Intelligence: Why it can Matter more than IQ* (Bantam Books, 1997)
Arthur Jensen: *The g Factor: The Science of Mental Ability* (Praeger, 1998)
Robert Plomin (with J.C. Defries, G.E. Mclearn and P. McGuffin): *Behavioural Genetics 4th edition* (Worth, New York, 2001)
Kevin Warwick: *IQ: The Quest for Intelligence* (Piatkus, 2000).

Big Questions in Science

How did Language Evolve

? Why did we begin talking to each other? At what point in human evolution did we need to say more than was possible with grunts, shrugs and an expressive pair of hands? And how is each new generation able to join in the conversation? Language is so deeply interwoven with how we think, who we think we are and how we understand each other, that for centuries philosophers and scientists have seen it as one of the cornerstones of human identity.

In Greek, Norse and Indian mythology, language was seen to be so special that it was attributed to divine intervention. In the biblical story of the Creation, the first man, Adam, is given the power by God to name what he sees around him in the Garden of Eden. But the first piece of rigorous thinking about the origin of language might be ascribed to Plato. In *Cratylus*, a Socratic debate on language, he examines whether there is an organic link between how a word sounds and what it represents or whether words are arbitrary labels attached to an object. Many later efforts to understand the origin of language have touched on this division. The challenge is that human language is of nature, but one step removed from instinct. While speech is an animal noise, which derives from our physiological development, it is also more than just a noise.

The idea that vocabulary has a natural root, and that the sound of a word is tied to its meaning, persisted for centuries. In the nineteenth century, many theories tried to find a bridge between human speech and sounds in the natural world, viewing language as a refinement of instinctive sounds. These theories included the so-called 'bow-wow' theory, which believed language derived from imitations of animal sounds, the 'pooh-pooh' theory, which claimed that words had begun as innate emotional sounds, such as for anger or happiness, the 'ding-dong' theory, which suggested that a word such as 'mama' was the sound made by a child's mouth seeking to suckle, and the 'yo-he-ho' theory, which saw language as the product of the kind of chanting and repetition that accompanied communal work. Eventually, in 1866, the foremost linguistic forum of its day, the Linguistic Society of Paris, decided that enough was enough: there should be no more debates on the origin of language because it could never be proved and because so many eccentric theories were making the subject ridiculous.

But the subject did not go away. Instead, different critical and scientific tools, drawn from fields including evolutionary biology, linguistics, genetics and anthropology, were applied to thinking about the mechanics of language and the way that it might first have been acquired.

At the end of the nineteenth century, the Swiss linguist Ferdinand de Saussure asserted that language was a system of arbitrary labels, so laying the foundations for what would become the structuralist school of linguistics. He argued that the only meaning a word had was that given to it by society. According to this view, the only way to understand an individual unit of language was in its relation to other parts of language. It meant that all vocabularies, regardless of where they occurred, were cultural phenomena and not innate, although it did not tackle where each of these different tongues had originated.

It had been established as far back as the eighteenth century that there were many connections between apparently different languages and that all Indo-European languages had a shared ancestry, with links between languages as diverse as English, Italian, Welsh, Swedish and Sanskrit. But, from the 1950s, Noam Chomsky's research at the Massachusetts Institute of Technology took this much further, analysing the underlying structures that were common to all languages and identifying a 'universal grammar'. He found that, regardless of culture or location, children share a facility for imitating and acquiring language. While a child in France speaks French and a child in Japan speaks Japanese, the same fundamentals and constraints are in place for language development in both. Even though language does not spontaneously emerge – a child without human contact has no speech – Chomsky saw an innate facility for language that was a physical rather than cultural property.

This opened the door to looking for the origins of language not only in language itself but also in the biological history of its speakers. So linguistics researchers such as Steven Pinker, MIT professor of psychology, began to look at language as another aspect of human evolution, which developed alongside physiological changes and which could be traced back through the human family tree. Instead of considering language a separate cultural attribute, it placed it within the 'out of Africa' theory of human evolution, which suggests language originated in East Africa around 100,000 to 150,000 years ago, evolving from a 'proto-language' which might have been used for 100,000 years before that. Voices leave no fossils, so dating depends on identifying the factors that could have driven language acquisition.

Jean Aitchison, professor of language and communication at Oxford University, says several different circumstances united to provide the needs and means for language: climate change had left our ancestors in an arid savannah, under pressure and needing to cooperate to survive; the human brain had grown – perhaps because of eating more meat as vegetation became scarce; our upright stance had enabled us to make a wider variety of sounds; and bigger brainpower allowed us to control mouth, lips and tongue and other muscles needed for speech. Intellectual preconditions for language, such as being able to consider other people's perspectives, the capacity to deceive and the ability to apply and recognise names for objects, also contributed.

Professor Aitchison says that once language emerged it would have become a key survival tool as our human ancestors left Africa. Different groups of early humans would probably have used different words, but dialects and vocabulary would have been subject to their own evolutionary power struggles and the stronger groups might have asserted their form of language over the weaker. Or else those humans who had a more advanced form of language might have used this organisational advantage to dominate those who had remained with only the rudiments of speech.

This model of events sees language not as a sudden thunder flash of intellectual advancement, but as an incremental separation from other primates. And this gap between human and non-human sounds would have been widened by physical changes, such as the lowering of the human larynx. This makes us susceptible to choking, but as an evolutionary pay-off, it makes it easier to produce a more differentiated range of speech sounds.

The pace of change of human language, from first steps to full development, remains uncertain. John Locke, a biolinguist at Cambridge University, says that there could have been a long interval between the first stages of language, when people only had a 'small repertoire of words', and the development of a wide vocabulary and grammatical structure. Between the emergence of fully-fledged language 100,000 years ago and the first incontrovertible signs of its use – in the form of writing – there is huge scope for speculation with little definite evidence. It is likely that language was relatively advanced when human communities were organised enough for an activity such as farming, but that still leaves tens of thousands of years in which the level of diversification of language remains unknown. Locke also points to unexplained curiosities, such as the way that certain kinship words, such as mama- and dada-type parent words, seem to cross language boundaries, even though this runs counter to all expectations about the arbitrariness of naming.

Another aspect of the origins of language not fully understood involves links between speech and the propensity to be left- or right-handed. Most people usually have the specialised language functions in the left hemisphere of the brain but a larger proportion of left-handed people appear to have them in the right hemisphere. John Coleman, director of the phonetics laboratory at Oxford University, says that investigating development of handedness could be a way of exploring the roots of language. Michael Corballis, professor in psychology at the University of Auckland, further explores the link between hands and speech, suggesting that the precursor of human language was not sound-based but was in a form of sign language based on hand gestures.

Genetics are also providing ways of examining language evolution. The Wellcome Trust Centre for Human Genetics in Oxford has used information from the Human Genome Project to identify a specific gene that produces proteins governing other genes associated with speech. The next stage of this research will be to find similarities and differences between this genetic code and the genetic code of other related primates in an attempt to isolate a genetic origin for language.

Tim Crow, director of the department of psychiatry at Oxford University, has suggested that a single genetic mutation in the Y chromosome in a human ancestor could have kick-started the process of language acquisition. According to this theory, humans who inherited this mutation would have had greater language skills, which would have given them an evolutionary advantage over other humans, leading swiftly to the dominance of those who had language.

In *The Mating Mind*, Geoffrey Miller suggests that more gifted linguists spread their genes more successfully because language skills can be a way of attracting sexual partners. This builds on other research into language as a social tool, including the theory that language developed as a much more efficient way of relationship-building among large groups than the kind of physical grooming common among primates.

Mario Vaneechoutte, microbiologist at the University of Ghent and editor of the *Journal of Memetics*, has also put forward the 'musical primate' theory. This claims that language has developed from song and musicality, with the ability to copy sound patterns being developed into a gradually more complicated system of symbols and meanings.

But any search for the origin of language will always have to acknowledge its unique place in human experience, says Michael Studdert-Kennedy, professor emeritus of linguistics at Yale University. Language is subjective, depending on the perceptions of both speaker and listener. And when we seek to analyse language, the tool we have to use is language itself. Language is not just a way of exchanging thoughts, says Studdert-Kennedy. It is the instrument of thought, shaping as well as expressing how we think.

Sean Coughlan
Freelance writer

How did language evolve?

Geoffrey Miller

Evolutionary psychologist at the University of New Mexico

We can speak; chimps cannot. Why? Explaining language remains the Big Question in human evolution, and a key challenge in my field of evolutionary psychology. Yet the more we learn about animal communication, the more mysterious human language looks.

Twenty-five years ago, language seemed easier to explain. In the late 1960s, John Pfeiffer argued in *The Emergence of Man* that language must have evolved with the 'Upper Paleolithic revolution' – the sudden appearance of cave art, carved figurines, burial rites and more complex tools in Europe 40,000 years ago. In the early 1970s, linguistic science professor Philip Lieberman claimed that Neanderthals could not have spoken, given the fossil evidence about their throat anatomy. And animal behaviour researchers such as Konrad Lorenz still had a naive view that many animals communicate to share useful information about the world.

This made a tidy story: language did not evolve at all in any other species of human-like primate; it only evolved in our species 40,000 years ago; and it evolved to share knowledge within groups. Once it evolved, we quickly invented culture and civilisation.

The problem is that, in the light of new evidence, none of these arguments work any more. If language evolved 40,000 years ago in Europe, how do we explain the fact that Africans and Australian Aborigines can also speak – given the genetic evidence that they diverged from Europeans at least 40,000 years ago? Psychologist Steven Pinker showed in *The Language Instinct* that language is a universal part of human nature, and since humans evolved at least 100,000 years ago in Africa, language must be at least that old. Palaeontologists have also overturned Lieberman's claims about mute Neanderthals. At most, the fossils suggest they might not have been able to produce the whole range of vowel sounds that modern humans do. That does not mean they could not speak.

Most importantly, British zoologists Richard Dawkins and John Krebs revolutionised the study of animal communication in 1978. They argued that it would be very odd for animals to evolve ways of giving away useful information to their evolutionary rivals. Communication in that sense would be altruistic, and it is very difficult for altruistic behaviours to evolve.

Since the Dawkins–Krebs revolution, biologists have discovered that most signals that animals send to each other are not messages about the world, but messages about the signaller. Many animal signals simply reveal the signaller's species, sex, age or location. Others reveal the signaller's needs, as when baby birds beg with open mouths to advertise their hunger to their parents. Most common of all are signals that reveal the signaller's fitness – their health, energy level, good brains or good genes – to deter predators from chasing them, to deter rivals from fighting them, or to attract sexual partners who are seeking fit mates. Many animal signals, from birdsong to whale song, from fruit-fly dances to the voltage surges from electric fish, say nothing more than: 'I'm here, I'm male, I'm healthy, copulate with me.' The signal's form may be complex, but its message is simple. A few social insects, such as bees, inform their sisters about food sources; a few mammals warn their relatives about dangers from predators. But even these signals about food and predators are simple, stereotyped and lazy – the bare minimum necessary to help the survival of their blood kin. Otherwise, most animals keep their knowledge about the world, quite selfishly, to themselves.

This makes human language look puzzling from a Darwinian viewpoint. Why do we bother to say anything remotely true, interesting or relevant to anybody who is not closely related to us? In answering this question, we have to play by the evolutionary rules. We cannot just say language is for the good of the group or the species. No trait in any other species has even been shown to be for the benefit of unrelated group members. Nor can we say language just popped up because of a single big mutation. If speaking is altruistic, then that mutation for speaking would have been eliminated very quickly by selection.

The evidence from psychology, linguistics and genetics shows that human language is a complex biological adaptation, and adaptations can only evolve gradually, over thousands of generations. They evolve because their evolutionary benefits consistently outweigh their costs. The evolutionary cost for language was telling useful things to non-relatives, which would allow their genes to prosper at the expense of one's own genes. But what were the survival or reproductive benefits of speaking?

Most popular books on language ignore the altruism problem and do not identify any specific evolutionary benefits of speaking. This is the weakness of *The Language Instinct*, Jean Aitchison's *The Seeds of Speech*, Derek Bickerton's *Language and Human Behaviour* and Terence Deacon's *The Symbolic Species*. This is also the weakness of so-called 'ape-language research'. Chimps only learn visual symbols when human experimenters, such as Sue Savage-Rumbaugh, bribe them with food to do so. Where were the beneficent experimenters who rewarded our ancestors for speaking on the African savannah 200,000 years ago?

British evolutionary psychologist Robin Dunbar developed one of the few theories that solves the altruism problem. In *Grooming, Gossip, and the Evolution of Language*, he argues that language evolved as an extension of primate grooming behaviour. Social primates maintain their relationships with other group members by grooming each other, up to several hours per day. Dunbar points out that as group sizes increased during human evolution, the time-costs of grooming would have increased to unsustainable levels. Perhaps language, especially social gossip, evolved as a more efficient way of servicing our relationships. The social benefits would have translated into both survival and reproductive pay-offs – as good relationships do – in primate social groups.

The trouble is, Dunbar's theory fails to explain why language has content. Why could we not have serviced our relationships by singing meaningless tunes to each other – like the 'signature whistles' of dolphins, or the 'contact calls' between primates? Dunbar jokes that his theory explains why most of our gossip seems so vacuous – 'Nice weather', 'Did you see how much weight Geri lost?' Yet what we consider trite any other species would consider astoundingly rich in meaning. If language is just verbal grooming, why is it about anything?

To solve the altruism problem and to explain why language has content, I think we need to update a theory proposed by anthropologist Robbins Burling in 1986. Burling noted that men in every society get social status for their public-speaking ability, and social status translates into reproductive success by attracting women. So perhaps language evolved through sexual selection, just like birdsong, with females favouring the best male orators. Bill gives good speeches, so Monica falls in love with him. He would have had extra babies under prehistoric conditions, and she would have benefited by merging her genes with his good-language genes to produce silver-tongued offspring. Thus there would have been runaway sexual selection for male language ability, and for female abilities to understand and judge language.

One problem with Burling's theory is that it does not explain why women talk too. Most sexually selected signals appear only in males because, in most species, males do all the courting and females do all the choosing. Female birds and whales do not sing; why do women speak if language evolved through sexual selection?

Unlike most other primates, humans form long-term sexual relationships, and mostly have babies within relationships (although there is plenty of infidelity). Since male humans invest more in their relationships and their children than any other primate, they have more incentives to be choosy about their long-term sexual partners. If our male ancestors favoured verbally fluent females over inarticulate or boring females, then sexual selection would have shaped female language abilities as well as male language abilities. The mutuality of human mate choice was crucial in giving us sexual equality in our adult language abilities.

Burling's theory also has the same trouble explaining content as Dunbar's theory. I think this problem can be solved by thinking about what a big-brained species would want to advertise during sexual courtship. If intelligence is important for survival and social life, then it would be a good idea to choose sexual partners for their intelligence. Language makes a particularly good intelligence indicator precisely because it has rich content. We put our thoughts and feelings into words, so when we talk to a potential mate they can assess our thoughts and feelings. We can read each other's minds through language, so we can choose mates for their minds, not just their bodies or songs. No other species can do this.

Language evolved because our ancestors favoured sexual partners who could show off what they knew, remembered and imagined. The prehistoric Cyranos reproduced more successfully than the Homer Simpsons; likewise the prehistoric Scheherazades. They did not always speak the truth about the world, but their language abilities always told the truth about themselves – the qualities of their minds and personalities that really mattered when sustaining relationships and raising children together. Language is not used only for verbal courtship. Yet I suspect that the origin of language lies in the way that our ancestors fell in love.

Further reading

Jean Aitchison: *The Seeds of Speech: Language Origin and Evolution* (Cambridge University Press, 2000)
Noam Chomsky: *On Language: Chomsky's Classic*

Works *Language and Responsibility and Reflections on Language in One Volume* (New Press, 1998)

Robin Dunbar: *Grooming, Gossip and the Evolution of Language* (Faber & Faber, 1997)

Geoffrey Miller: *The Mating Mind* (Heinemann, 2000)

Steven Pinker: *The Language Instinct: How the Mind Creates Language* (HarperCollins and Penguin, 1995).

Big Questions in Science

Are We Shaped By Nature or By the Environment ?

The sequencing of the human genome may have sparked a new interest in the interplay between genes and environment, but the nature/nurture debate has been running for centuries. 'Nature has always had more power than education,' said Voltaire in the 1700s, revealing where he stood. His contemporary, Jean-Jacques Rousseau, on the other hand, believed that men were born good but corrupted by the society in which they lived. Earlier, René Descartes had written of inner drives in conflict with conscious reason. By contrast, English philosophers John Locke in the seventeenth century and John Stuart Mill in the nineteenth century thought that behaviour derived from observing the world, not from some innate drive.

Only in the past 150 years has it been possible to tackle the question in anything resembling scientific terms. It was Gregor Mendel who laid the foundations for the study of inheritance and Charles Darwin who described a mechanism of selection by which organisms with the most favourable inherited characteristics would be more likely to pass on those characteristics to future generations.

Mendel was ordained a priest but studied science in Vienna before returning to the monastery at Brunn of which he eventually became abbot. His scientific interest was in plant breeding – in the hybrids born of crossing plants with distinctly different characteristics. His classic experiments involved the colour and form of pea seeds, which could be round or wrinkled, yellow or green. Mendel unravelled the pattern of their inheritance, showed that some characteristics were dominant and some recessive, and, crucially, demonstrated the arithmetic proportions in which these characteristics would occur.

It was clear that inheritance involved the passage from one generation to another of discrete entities of some kind – packets of information. Mendel did not comprehend the nature of these packets, or genes, as we now refer to them. Nor did Darwin, who also remained ignorant of Mendel's work on peas. But not knowing exactly how specific characteristics were passed from generation to generation did not prevent Darwin from devising his theory of natural selection.

His ideas derived from a series of observations and premises. Offspring resemble their parents – but not exactly. Given that all organisms produce more offspring than can hope to reach maturity, the more successful of these variants, he reasoned, stand the best chance of surviving to breed. This ensures the transmission of favourable characteristics to the next generation. Over time, these will tend to become more common in the population, so improving an organism's adaptation to its circumstances.

That this process accounts for the evolutionary changes which have led to such a bewildering variety of form and function in the living world is accepted by virtually all biologists. Much the same is true for most animal behaviour. The sticking point is human behaviour. We inherit our sense organs, our muscles and our brains; but do we also inherit the intelligence needed to use them in building a computer, the determination required to make it work, or the generosity to lend it to someone else?

Confronted with questions of this kind, Europeans and Americans have put differing emphases on their answers. In Europe, nineteenth-century English philosopher Herbert Spencer seized on Darwin's *Origin of Species* to invent a form of social Darwinism that sought to justify every feature of unbridled capitalism by opposing intervention by the state. And throughout the first half of the twentieth century, nature rather than nurture continued to be viewed as the predominant influence. One of its leading advocates was Darwin's cousin, Francis Galton. He championed a fledgling eugenics movement devoted to programmes of selective breeding that would improve the human stock. Radical intellectuals such as George Bernard Shaw and Sidney and Beatrice Webb were attracted to these views, although Hitler and the Third Reich's support for eugenics soon made them unacceptable.

Sigmund Freud, too, accepted the existence of certain broad innate drives, but he then stressed the importance of how the events of early life influenced and shaped them. His school of psychoanalytic thinking, along with those of its contenders, found readier acceptance in America than in its European birthplace.

For much of the first half of the twentieth century, psychology in America was strongly influenced – if not dominated – by behaviourism, which holds that there is little innate programming of the newborn brain, that we react in response to certain stimuli, and that our behaviour is the product of our conditioning. The psychologists J.B. Watson and, subsequently, B.F. Skinner emerged as the leading champions of this view. There were parallel developments in the social sciences, with cultural anthropologist Margaret Mead and others viewing the human mind as a tabula rasa, and human society as therefore highly malleable. Culture rather than biology was in the ascendant.

Rebellion was inevitable. One of the most charged areas of debate concerned the inheritance of IQ, particularly in connection with race. The academics Arthur Jensen in America and Hans Eysenck in Britain emerged as hate figures for their highly publicised beliefs that much in our character and behaviour is inherited. Eysenck, for example, liked to claim that IQ was shaped by nature and nurture in the ratio of 4:1.

A more coherent, more solidly Darwinian view of the nature/nurture debate began to re-emerge in the last three decades of the century, partly through recognition of the painstaking studies of animal behaviourists such as Konrad Lorenz, Niko Tinbergen and, above all, E.O. Wilson. It was Wilson, in his 1975 epoch-making book *Sociobiology: The New Synthesis*, who was most explicit in resurrecting the possibility that the roots of human behaviour might be genetically determined, and that distinctions between humans and animals in this respect were misguided.

Wilson's views created a storm, intellectually and politically. He was accused of racism and sexism, and dismissed as a biological determinist who held that genes were destiny. The term sociobiology became so discredited that even those sympathetic to Wilson's thinking were reluctant to use it.

Meanwhile, a group that included not just biologists but philosophers and social scientists had gone back to Darwin and founded a new school of evolutionary studies. In broad terms, the claim of this school is that, through natural selection, our species became adapted to its long period of living a hunter-gatherer existence. The environment and social organisation prevailing at that time were markedly different from today's. But, biologically speaking, we are now as we were then because, as a force for changing humans, biological evolution has been largely superseded by cultural evolution. So the new programmes of learned human behaviour have been grafted on to old inherited ones. If you want to know what comes 'naturally' to our species, so the argument goes, ask how we lived some tens of thousands of years ago.

When applied to physiology this has generated lively debate and intriguing insights on matters such as the prevalence of obesity and the rise of heart disease. But the issues have remained largely non-controversial. The consequence of applying the same principle to human behaviour – evolutionary psychology, as it is known – has generated a rerun of the rows over sociobiology, the discipline from which it is clearly descended. The arguments about evolutionary psychology have been slightly less bitter, but no less heated.

The biologist Steven Rose, a fierce critic of evolutionary psychology, calls the emerging synthesis of brain science and genetics 'neurogenetics', and describes its alleged effects on human behaviour as 'neurogenetic determinism'. In *Lifelines*, he writes: 'While only the most extreme reductionist would look for the origins of the Bosnian war in deficiencies in neurotransmitter mechanisms in Dr Radovan Karadzic's brain, and its cure the mass prescription of Prozac, many of the arguments offered by neurogenetic determinism are not far removed from such extremes.'

On the other side of the argument, Wilson himself denies saying that genes are destiny, and admits he was naive in failing to anticipate the fury that sociobiology would provoke. He does, though, claim to understand the continuing reluctance of many people to accept ancestry as a factor in determining behaviour, saying: 'It is a more comfortable as well as a more easily comprehended position that all minds start exactly the same, all potential is the same, and that one only needs to change the environment to change human behaviour in any direction.'

To what extent the findings of the Human Genome Project will sway the debate is hard to say. Identification of a load of genes does not, in itself, alter anything. But as researchers work their way through them, determining their various functions, the picture could alter. Indeed, many in the 'nature camp' believe that the wind is now blowing firmly in their direction. What started with the outcome of twin studies will, they believe, find confirmation in molecular biology.

Many of them are nevertheless keen to avoid the fruitless polarisation of unbridled genetic determinism versus unbridled cultural determinism. Most scientists now accept that both genes and culture play a role. The focus of the debate has shifted to their relative rather than their absolute influence.

Geoff Watts
Writer and broadcaster on science and medicine

Are we shaped by nature or by the environment?

Michael Rutter

Professor of developmental psychopathology,
Institute of Psychiatry, Kings College, London

Until recently, behaviour genetics was mainly concerned with quantifying the relative influences of nature and nurture on psychological development and mental disorders. Twin and adoptee studies were primarily used to separate genetic and environmental effects. The findings were consistent and important in showing the strengths of both nature and nurture. Overall, their effects appeared to be approximately equal, although genetic factors clearly predominated for some disorders (such as autism or schizophrenia), whereas environmental factors did so for others (such as crime).

Now it has become clear that to regard nature and nurture as separate and independent is a misleading oversimplification. Effects depend on interplay between the two, both in terms of correlations and interactions.

Correlations arise because genes influence individual differences in exposure to environmental risks through three different mechanisms. First, parents both pass on their genes to their offspring and provide their children's upbringing environment. The correlation between genetic and environmental influences reflects the fact that, on the whole, the parents who pass on genes involving an increased risk for mental disorders also tend to provide less than optimal rearing environments. For example, parents who are seriously troubled by their own recurrent depression, or persistent drug or alcohol problems, may find parenting more difficult. The risks to their children are thus a combination of genetic and environmental. In traditional analyses, the combined effect is attributed entirely to genetics whereas, in reality, it involves the joint action of nature and nurture.

Second, people select and shape their environments through their own behaviour. Thus, for example, a child with genetically influenced musical, athletic or mathematical talents is likely to spend more time (and possibly higher quality time) in these pursuits than other children. Further development of such talents will therefore be influenced by these environmental advantages, as well as by the child's genetic background. The genes will have played a key role in shaping and selecting the environments, but the influence will reflect the coming together of nature and nurture. Again, traditional analyses will attribute all the effects to genetics, despite the environment's mediating role.

Third, people's genetically influenced behaviour affects their interactions with others. For example, antisocial individuals are more likely than others to act in ways that provoke hostility or rejection, lead to a lack of social support, predispose to the breakdown of relationships and put their jobs in jeopardy. All of these effects involve substantial environmental risks. Once more, genes are important in making it more (or less) likely that individuals will experience risky environments, but although ordinarily included in the estimate of genetic effects, the risks actually involve both genetic and environmental mediation.

These findings carry crucial implications for both genetic and psychosocial research. The message for genetics is that part of the genetic effect lies in its indirect impact on variations in exposure to environmental risks. It therefore involves both nature and nurture, and labelling it as just genetic is misleading. The message for psychosocial research is a parallel one: some of the effects that appear to be wholly environmental are, in reality, partly genetically mediated.

Genetic evangelists have sometimes sought to rubbish psychosocial research on this basis. But their criticisms are unwarranted, firstly because the genetic findings show that, usually, only a minority of the supposed environmental effects are genetically mediated, and secondly because genetic analyses have confirmed that environmental risk mediation exists. For example, environmental factors have been shown to account for the differences in outcome within monozygotic (identical) twin pairs who share all their genes.

So much for the correlation between genes and environment. Gene–environment interaction reflects a rather different mechanism. It has been a universal finding in environmental risk research that children (and adults) vary enormously in their responses. For any given environmental hazard, however severe, some individuals suffer greatly and some seem to escape the main adverse outcomes. Genetic factors play a crucial role in that individual variation in susceptibility or vulnerability. This kind of effect applies across the whole of biology and medicine. Thus, exposure to pollens in the spring gives rise to severe hay fever in some individuals, whereas others are completely unaffected; genetic influences are concerned in that individual difference. Moreover, molecular genetic research, which studies the effects of individual susceptibility genes, has confirmed that genes and environment work together in relation to risk factors as varied as smoking, head injury and infections. Such disorders are unlikely to occur in the absence of susceptibility genes, and are unlikely, too, in the absence of the environmental risk factor. It is the presence of the two that is crucial. Once more, old-style quantitative genetic analyses will have attributed the whole of this effect to genes, whereas it actually stems from the combination of nature and nurture.

The existence, and pervasiveness, of gene–environment correlations and interactions mean that any assessment of effects needs (at a minimum) to deal with nature, with nurture, and with the combined effect of the two. Too few data exist for any general conclusion on the relative size of this combined effect and it is likely to vary according to different traits or disorders. Clearly, it is not trivial, but it is also necessary to avoid overstating its importance. We still need to ask whether there are major genetic effects that are independent of environmental adversities and whether there are environmental effects on individuals who are not genetically susceptible.

The independent importance of genetic effects has the strongest factual basis. Evidence concerning both schizophrenia and autism, for example, indicates that the genetic risks for these disorders are not dependent on the children encountering environmental hazards of any type. The same probably applies, to some degree, to other psychological features. By contrast, on the whole, environmental effects are most evident in genetically susceptible individuals. Some environmental effects probably exist that do not require genetic susceptibility, but they have not been so clearly demonstrated.

Two other caveats have to be added to the nature/nurture question. First, non-genetic influences do not necessarily involve specific environmental effects. That is because biological development is probabilistic rather than deterministic. In other words, the evolutionarily derived genetic programme specifies a general pattern or plan, but it does not determine what each individual nerve cell (or any other type of cell) does. Chance and general perturbations play a considerable role. Thus, every female has two X chromosomes but only one is active and which one seems to be largely determined by chance. Whichever it is will matter in some circumstances because one X is inherited from the father and one from the mother. General perturbations are widespread in development. Thus most of us have minor anomalies of one kind or another – an extra nipple, an extra tooth, a missing muscle, an unusual eyefold, an asymmetric skin pattern or unusually constructed ears, for example. Such anomalies are meaningful at a group level – they are more common in twins than singletons and in children born to older mothers – but no specific environmental factor seems responsible for their occurrence at an individual level. Moreover, most anomalies have no functional consequences. Nevertheless, they may be important because they point to development going slightly awry. Disorders may result from some combination of genetic risk and developmental imperfections rather than from any particular environmental risk experience.

The second caveat is that although quantifying genetic and environmental effects looks at individual differences, we also need to consider their effects on the frequency of a particular trait. Over the last half-century, there have been huge rises in the rates of substance abuse and crime among young people and in suicide among young males. The speed of these rises clearly points to an environmental effect of some kind. Over the course of the twentieth century, there were also rises in height and IQ, and a fall in the age of menarche. Again, environmental factors will have been responsible. The pattern of research findings indicates that the factors responsible for individual differences in a particular characteristic are not necessarily synonymous with the factors responsible for the level or frequency of that characteristic in the population as a whole. Thus genetic factors are largely responsible for individual differences in height, but the huge rise in average height (some twelve centimetres or so) over the last century is almost certainly due to improved nutrition. A high, even a very high, heritability does not mean that a major change in environmental circumstances cannot make a big difference.

Where does that leave the question of whether we are shaped by nature or the environment? The answer has to be both. However, research findings go further than that in emphasising the interplay between the two. Much of the variation between people stems from the synergistic combination of nature and nurture.

In a sense, the question is the wrong one. There are no policy or practice implications in heritability being high or low (except, perhaps, when it is near 0 per cent or 100 per cent). What really matters is not the relative strength of genetic and environmental effects (which will in any case vary according to circumstances) but the mechanisms by which they exert their effects. Therein lies the future. Do genetic influences on antisocial behaviour operate via the indirect risks associated with sensation seeking or impulsivity, the more direct risks associated with aggressivity, or the protective effects associated with high anxiety?

Molecular geneticists will play a crucial role in providing that understanding of causal processes. Until now, most biological research in the field of mental disorders has been inconclusive because so much has had the nature of a fishing expedition. But once molecular genetics has identified one or more relevant susceptibility genes, and once functional genomics research has gone on to show the effects of these genes on proteins and on the biological processes the protein products bring about, it should help narrow down the search for underlying biological causal mechanisms.

However, research will only be fully successful if it includes investigation of the interplay between nature and nurture. This is because some crucial genetic influences on outcomes concern their effect on exposure and susceptibility to specific environmental risks. So research needs to extend beyond the processes operating within the cell, to the processes involved in how individuals interact with their environments, and hence to the indirect pathways by which genetically influenced liabilities lead to particular behaviours. This task can be accomplished, but success will not be easy and is likely to take a long time.

Further reading

Daniel Dennett: *Darwin's Dangerous Idea* (Allen Lane, 1995)

Steven Rose: *Lifelines* (Penguin, 1997)

Michael Rutter (with J. Silberg): 'Gene–environment interplay in relation to emotional and behavioural disturbance' (*Annual Review of Psychology* 2002, vol. 53, pp. 463–90), 'Nature, Nurture and Development: From Evangelism Through Science Towards Policy and Practice' (*Child Development*, vol. 73, no. 1,

Jan/Feb 2002), *Developing Minds: Challenge and Continuity Across the Lifespan* (Penguin, 1993)

E.O. Wilson: *Sociobiology: The New Synthesis* (Harvard University Press, 1975).

How are Men and Women Different ?

? 'Why can't a woman be more like a man?' Henry Higgins demands in the musical *My Fair Lady*, to the instinctive sympathy of his fellows. After all, men are so much better at the important things in life. They know the virtue of silence. They run with grace, throw a ball with accuracy, park a vehicle with perfection at first try. They don't cry, or take offence over trivial issues.

Over the last decade scientists have been adding to the list of male talents but also compiling a formidable index of female accomplishments. Women, as Higgins's protégée Eliza Doolittle amply demonstrates, are the more articulate sex – fluent and expressive, they are mistresses of grammar and superior at reading. They are more subtle creatures, capable of judging the moods and intentions of others with finesse and running a web of complex social relationships.

Humorists and philosophers have written of the sex differences for centuries, most believing that they described the immutable. It is only in the twentieth century that the belief that society moulds its own stereotypes gained real power. Applied to gender, the idea took grip in the 1960s as social scientists revealed the degree to which women's behaviour is dictated by society, in particular role models and the media. Many social scientists discounted any role whatsoever for biology in the creation of gender differences (apart from differences in reproductive machinery). They now stand accused of dogmatism by a biology-based discipline that is trying to reclaim some territory.

The neo-Darwinists have to tread cautiously. Their predecessors – social Darwinists – used Darwin's theory of natural selection to provide an apparently scientific basis for slavery, racial discrimination and conquest. Today they call themselves evolutionary psychologists, and their position on gender differences is this: men and women, driven by their distinct biologies to adopt different strategies for reproduction, have evolved different behaviours, values and ways of perceiving the world.

It all starts in the womb with a great rinsing of hormones that trigger, just twelve weeks after conception, the diverging of ways. In the uterus, the brain of the early foetus is innately female until this moment; then boy foetuses secrete male hormones known as androgens. As these hormones surge into the brain, they shape and re-organise it, emphasising some neural networks and suppressing others. Maleness has begun.

A window on to this process has been provided by the technique of amniocentesis. By withdrawing a little fluid from the womb, scientists can measure the exposure of an individual foetus to androgens. They can then try to correlate their behaviour as babies with their early hormonal environments.

Cambridge scientists Svetlana Lutchmaya and Simon Baron-Cohen have done just such a study. They selected a behaviour thought to be of major importance in normal social development: making eye contact with other humans. Evolutionary theory predicts that infant girls will engage in more eye contact than boys because of a deeper fascination with faces and emotions that is a prelude to developing greater social skills.

They found that, at twelve months old, those babies who had experienced the weakest androgen environment in the womb were capable of the greatest amount of eye contact with their parents. These superior communicators were generally girls, but there was overlap between the sexes, reflecting overlap in the amount of androgen exposure in the womb.

Findings such as these may not directly indicate that girls are born with better social skills. Instead, it is possible that boys and girls are born with gender-dependent interests, motivations to seek out different experiences. Girls, more interested almost from day one in observing expressions and emotions, gradually become more socially adept. This may be reflected in permanent brain changes.

'Early exposure to sex hormones leads to differences in brain organization that might not be very large but at the same time appear to bias boys and girls to engage in activities that will gradually increase these initial differences,' says David Geary, professor of psychology at the University of Missouri, Columbia, and author of *Male, Female, The Evolution of Human Sex Differences.*

Studies of three-year-olds find further differences. For some girls, the rinsing in the womb goes awry and, as a result of the condition known as congenital adrenal hyperplasia (CAH), they are exposed to unusually high levels of androgens. Compared with unaffected girls, CAH girls are more competitive athletically, according to Melissa Hines, psychologist at City University, London. They prefer mechanical and structural toys to dolls, rough-and-tumble play to play-caring for infants.

By the age of between eight and eleven, there is further

evidence of a parting of ways and the development of spatial skills in males. Boys of that age roam more widely than girls do. This is partly because they are given more freedom by their parents, but studies have tried to exclude this influence. They have found that although most girls are less keen to explore widely, they improve their navigation skills if they are encouraged to roam. Again, the different environmental experiences that boys instinctively seek may polish certain brain circuits and diminish others.

The hormonal effect continues into adulthood. Researchers have found that women's language abilities fluctuate with their oestrogen cycles, while men who change sex improve in some measures of language ability after hormone treatment.

The link between hormone levels and behaviour has been easier to demonstrate than that between brain structure and behaviour. Some scientists believe they have found structural differences between male and female brains. Godfrey Pearlson, of Johns Hopkins University, for example, believes that, through magnetic resonance imaging scans, he has found a brain region in the cortex, called the inferior-parietal lobule, that is significantly larger in men than in women. But very few such findings have been verified. And the work done so far to correlate such differences with behaviour is 'very crude', says Roger Gorski, neurobiologist at the University of California, Los Angeles. 'It's very, very difficult; the brain has many redundant control mechanisms.'

Scientists do seem to have found consistent differences in brain activity. Perhaps the most well-known example comes from the imaging of male and female brains while they do language tasks. Most women use both the right and left hemispheres of the brain, while most men use just one hemisphere. It is thought that the use of two hemispheres allows flexibility, fluency and the ability to perceive connections more easily.

For those who accept that the differing behaviour of girls and boys does have a biological component, the next step is to understand why. Evolutionary psychologists find an answer in sexual selection. This is the second great mechanism of Darwinian evolution, the first being survival of the fittest. It is not survival alone that guarantees that a gene will be passed to the next generation – reproductive success is also essential. Thus, genes that enhance reproductive success are selected as the generations go by. Crucially, genes that confer reproductive success on women will differ from those that confer such success on men. This is because of the basic biological realities of reproduction. However hard she tries, a female can only produce about one child a year, while a man's reproductive possibilities are almost infinite. The two realities call for different mating strategies.

For most of the animal kingdom, including humans, methods of mating have two distinctive components: competing with members of your own sex for access to mates and/or choosing your mate from what is available. Evolutionary psychologists have thus settled on four principal mechanisms by which to understand how male/female differences arose: male–male competition and female choice (the most common pair), female–female competition and male choice. The degree to which each of these mechanisms dominates depends on the cultural setting.

In many species and, it is argued, human societies, competition for access to females is the dominant activity in a male's life. Success can make the difference between monopolising access to many females, on the one hand, and not producing a single son or daughter, on the other. Genes that help men achieve social dominance, for example through superior fighting ability or political skills, will flood the next generation.

Females, traditionally vulnerable during pregnancy and early motherhood, and investing heavily in the well-being of a smaller number of offspring, have more success if they select males who can provide them with material resources and social stability. Thus female discrimination brings certain male characteristics to the fore – in pre-industrial societies, success in hunting, for example, for which navigational abilities and athletic prowess were indispensable. According to a study by psychologist David Buss of thirty-seven countries worldwide, female tastes remain similar whether the society is industrial or pre-industrial, Stone Age, Western, Oriental or African.

The third mechanism, female–female competition over desirable mates, is thought to be primarily linguistic. Liverpool University psychologist Robin Dunbar has shown that the conversations of girls and women often exhibit the tendency to try to exclude females who are perceived as rivals – the roots, perhaps, of gossip and 'bitchiness'. Gossip about sexual infidelity can spoil a woman's marriage prospects.

One of the more sensational discoveries of recent years relates to the final force – male choice. Buss's study revealed that men all over the world like women to have the shape of an hourglass. More precisely, they like a waist-to-hip ratio (waist size divided by hip size) of 0.7. Evolutionary psychologists argue that this ratio, which means the waist is 30 per cent smaller than the hips, reflects a combination of health, youth and fecundity. A ratio of above 0.85, for example, puts women at risk of a number of physiological disorders and lessens their chance of conceiving.

Some of these insights and observations may seem crude, but it is important to understand that most evolutionary psychologists claim that this is only the beginning of understanding gender differences. Biologically based tendencies can be tuned upwards or downwards, redirected, counteracted or suppressed by the environment. Culture can sculpt the instinct for male competition into homicidal violence or the pursuit of MBAs. We differ from the opposite sex because evolution constrains us to be different and hormones modify the raw material. But what we make of that material depends on society.

Aisling Irwin
Award-winning science writer

How are men and women different?

Janet Radcliffe Richards
Reader in bioethics at University College London

Were traditionalists right about men and women after all? They always said the sexes were different, and used their ideas about those differences as the excuse – whenever excuses were needed – for the relegation of women to their subordinate sphere. If science seems now to be confirming these old beliefs, it is not surprising that there is widespread alarm and insistent denial among feminists and their sympathisers.

It was the scientific approach to human nature that first challenged the traditional view. The sexes might seem very different; but, as philosopher John Stuart Mill pointed out, the extent to which these appearances reflected natural differences was impossible to tell because men and women had always been in systematically different environments. And the case against tradition seemed to gather strength as this agnosticism was replaced by the conviction of social scientists that observed differences between the sexes were cultural constructs. Feminists adapted the term 'gender' to refer to such non-biological differences; and it is a sign of how essential the social construction view has become to feminism that the substitution of 'gender' for 'sex' seems to have become compulsory among the politically enlightened.

Still, no matter how widely accepted it may have become, the cultural-construction view of psychological difference has always been implausible. If human beings are regarded as belonging entirely to the natural world, and their emotions and intelligence as – however mysteriously – functions of matter rather than properties of separately infused souls, differences in mind must ultimately be connected with differences of body. It would be astonishing if the main systematic division within the species managed to avoid any correlation with differences of mind; and science, which has progressed a good deal since the time of Mill, is increasingly proving that it has not. We now have direct physiological evidence of mental and emotional differences between the sexes; and even more is coming from the new direction of approach provided by evolutionary psychology.

Evolutionary psychology tackles the problem of disentangling natural and cultural differences between the sexes by using our understanding of evolutionary processes to generate hypotheses about natural differences. Darwin himself recognised that as soon as evolution produced creatures with emotions and intelligence, these qualities would be as relevant to the evolutionary fate of their possessors as would anything else about them. And when the sexes – of almost any sentient species – are considered from this point of view, it becomes clear that males and females should be expected to have significantly different temperaments, just because their reproductive systems are so different.

A human female, reproducing flat out, can produce only about a child a year. A human male's reproductive potential is limited only – though of course entirely – by his ability to impregnate women. This alone suggests that different psychological characteristics would have been needed for evolutionary success. A woman need make no effort at all to have children; men, unless actively kept at bay, will see to that. No amount of sex can increase her reproductive potential, so she will do best, if her emotions lead her to maximise the quality of her offspring, by selecting mates who are of high genetic quality, and who also (preferably) have important resources they can be relied on to give to those offspring. A male, on the other hand, is playing for higher stakes, since he may have far more than his fair share of children, and also risking greater losses, since he may have none at all. If he is to succeed in the evolutionary competition, he will need quite different emotional characteristics from those of females.

Reasoning along these lines has produced dozens of hypotheses about differences between men and women. And the problem is that many of these, dismissed by recent feminism as culturally induced, are turning out to be ones that natural selection might be expected to have planted deep in the nature of the sexes.

Evolutionary reasoning has suggested, for instance, that women should be attracted to impressive, high-status males from whom they seek undivided support and commitment, and should be strongly devoted to the care of their children. Men should be competitive, adventurous and anxious to possess women and control their sexuality. They should prefer youth and beauty in women, but also be eager to grasp whatever sexual opportunities present themselves.

It begins to sound as though, after all its early promise as an ally of liberation, the science of human nature is leading straight back to the traditional view. So it is hardly surprising that in many quarters the whole project of evolutionary psychology is dismissed as politically motivated pseudo-science, and is met with accusations of genetic determinism, essentialism, gross oversimplification, insensitivity to variation and overlap, categorising, stereotyping and rampant sexism.

But in fact the new claims about sex differences, although they may sound like the old ones, are really quite different.

In the first place, although the new science seems to be confirming several traditional views about male and female nature, it must be stressed that it has produced nothing like a clean sweep. There is nothing whatever in evolutionary psychology to support, for instance, traditional views of lesser female intelligence, or of inability of females to fend for themselves and protect their offspring, or of relative weakness in any respect but upper-body strength.

There is also a more fundamental, and more subtle,

point. Even when modern claims about male and female differences of nature sound like traditional ones, they are not really the same because there has been a radical change in the whole idea of nature and what it means to understand the nature of something.

The kind of change at issue can be illustrated by the anti-feminist writing of the jurist James Fitzjames Stephen, one of Mill's most articulate contemporary critics. He started with the claim that men and women were different 'from the hair of their head to the soles of their feet', with 'men stronger than women in every shape'; and, on the basis of this, he defended the traditional subordination of women to men in marriage. His official argument to this conclusion was that the arrangement was for the protection of women. This was manifest nonsense: you cannot protect the weak by putting them into the legal power of the strong and making them weaker still. But in the background are hints of what really underlies his convictions. He talks of the need for institutions 'to clothe, protect and sustain society in the position which it naturally assumes'. He thinks that a wife should submit to her husband's judgement in the same way as the first lieutenant should to the judgement of the captain of a ship, and that she shows a 'base, unworthy, mutinous disposition' if she resents this. And he thinks that all this is for 'the common good of the two great divisions of mankind', which can no more have different interests than can different parts of the same body.

All this suggests that Stephen was working against the background of a deeply rooted, traditional view of the world as a naturally ordered whole, in which all was harmonious as long as everything stayed in its ordained place. If things went wrong, that was because of interference in, or rebellion against, the natural order of things. This idea comes in many different versions, of which the most familiar now is the religious view that sees order and complexity as underlain by intelligent design.

Against the background of such traditions, to understand the nature of something is to understand its proper place in the scheme of things, and to understand the nature of men and women is to know how they should live harmoniously together.

But Darwin's account of evolution by natural selection, which showed in principle how complexity could arise from simplicity without any design or intention, revealed a quite different world, in which there was no underlying moral order, and no natural harmony. In this world – as, indeed, in the world of modern science in general – to describe the nature of something is to say nothing at all about its natural place or what is good for it. It is only to give a neutral account of what it is like, and how it interacts with other things.

The trouble is, however, that pre-scientific, pre-Darwinian ideas about nature are deeply ingrained in our consciousness, and persist even among people who have, in theory, abandoned them. This leads to systematic misinterpretation – or misrepresentation – of the kinds of claims made by evolutionary psychology.

In a Darwinian world, for instance, claims about the way evolution has shaped male and female emotions carry no implications at all of psychological homogeneity within the sexes, or of firm boundaries between them – as critics often claim. Since variation is the raw material of evolution by natural selection, variation is to be expected. Ideas of fixed essences and clear distinctions between natural kinds belong only to earlier ideas of an ordered universe.

In a Darwinian world, similarly, claims about sex differences carry no implication of genetic determinism of any kind. To say that men and women are different by nature does not imply that their development and actions are fixed in their genes; it implies only that – to the extent that they are different – they will react in different ways to similar environments.

To understand the nature of something is, precisely, to understand the circumstances under which that nature will change. The idea that natures are unchanging also belongs to the pre-Darwinian world, and has nothing to do with the claims of evolutionary psychology.

And, most important of all, in a Darwinian world,

discoveries about the natures of the sexes have no direct implications whatever for how they should live and relate to each other. There is not the slightest reason to expect their interests – either evolutionary or personal – to coincide. Indeed, the evolutionary writer Robert Wright says the sexes almost seem designed to make each other miserable. Natural selection produces harmony only to the extent that harmony promotes reproduction. Evolutionarily speaking, the sexes are rivals. Whatever may be achieved by well-matched or well-motivated individuals, there is no natural prescription for domestic harmony or social justice. To think otherwise is once again to import incompatible, traditional ideas of an ordered universe into the world of modern science.

And, paradoxically, the same mistake shows even in the apparently opposed idea that Darwinism justifies constant struggle. An idea popular among some males of the species is that if their inclinations to philander or even rape are evolutionary adaptations, we should not impede evolutionary progress by trying to curtail them. But ideas of evolution as all-purpose progress also depend on the idea of a natural order, through which evolution progresses. Darwinian evolution has no such onward-and-upward path. The only hope for progress of any kind lies in our deciding what counts as progress, and then trying to bring it about.

Whatever we hope to do, we cannot do it without understanding what we are up against. If science is showing us that men and women are different, as it certainly seems to be, that is something we need to know. The last thing we can afford is resistance that comes from encumbering Darwinian claims about human nature with fossils from a pre-Darwinian world.

Further reading
David Geary: *Male, Female, The Evolution of Human Sex Differences* (American Psychological Association, 1998)

Big Questions in Science

Janet Radcliffe Richards: *Human Nature After Darwin* (Routledge, 2000), *The Sceptical Feminist* (Penguin, 1984)

Matt Ridley: *The Red Queen: Sex and the Evolution of Human Nature* (Penguin, 1995)

Steven and Hilary Rose (eds): *Alas Poor Darwin: Arguments Against Evolutionary Psychology* (Jonathan Cape, 2000).

What Makes Us Fall
In and Out of Love

?

We have all felt that little 'zing!' when we see someone we fancy, and a much greater 'ZING!' when these feelings are reciprocated. What is happening? The question of why we experience 'zings' can be answered at two levels: first by asking what makes someone attractive, and second by asking why we have evolved to find certain features attractive. The first is a question of immediate causes, the second addresses the evolutionary significance of attraction. Usually, these two types of question are tackled by different sorts of researchers, concentrating respectively on physiological mechanisms and evolutionary factors. But evolutionary biologists interested in sexual attraction have addressed them together.

Those evolutionary biologists who study animal behaviour are often confident that they know why a particular male and female have ended up copulating together. It is because either the female has chosen the male, the male has competed for and 'won' the female, or because both partners have come to some mutual agreement about liking each other.

It was Charles Darwin who first put sexual attraction into an evolutionary context. In formulating his ideas about natural selection, he was concerned about traits that did nothing to promote their owner's survival. The extravagant plumage of male birds and the cumbersome antlers of deer made their owners conspicuous and vulnerable to predators, so how could they possibly evolve by natural selection? Darwin's answer was sexual selection. Extravagant traits probably did reduce their owner's survival by making them more vulnerable to predators, but they more than compensated by making their owners more competitive or irresistibly attractive to members of the opposite sex, so that they left more offspring – more copies of their genes – than less ornamented males.

Sexual selection accounts for many of the differences between males and females and Darwin saw it operating through two processes: competition between members of the same sex – usually males competing for females – and choice by one sex for some members of the other – usually female choice of males. Competition between males for females accounted for the evolution of weapons such as teeth, spurs and antlers, while choice of males by females accounted for otherwise useless ornaments such as plumes, wattles and perfumes. Sexual selection is about differential reproductive success – the more attractive or competitive individuals are, the more descendants they will leave.

Darwin's contemporaries had no problem with the concept of male–male competition. They could it see it happening in every farmyard, and the idea coincided nicely with Victorian ideas about sex being driven by active males pursuing passive partners. Female choice was another matter altogether. It was invisible, and anyway, they (men) said, females probably did not have enough brainpower to make such informed choices.

When Darwin died in 1882 the idea of female choice all but died with him and, like sleeping beauty, it remained (virtually) unconscious for one hundred years. The person who planted the awakening intellectual kiss and roused the whole field of sexual selection to new heights was behavioural ecologist Malte Andersson of the University of Gothenburg in Sweden. He conducted an elegant experiment that involved shortening and elongating (using scissors and superglue) the tail feathers of male long-tailed widowbirds, and showed that females preferred the long-tailed males. Female choice was alive and flying and, with some more experimentation, subsequent researchers found female choice to be widespread throughout the animal kingdom.

Behavioural ecologists rediscovered Darwin in the early 1970s. It was a paradigm shift, and one that, like so many, started off with some sloppy thinking and much controversy. In 1981, for example, palaeontologist Stephen Jay Gould and evolutionary geneticist Richard Lewontin accused behavioural ecologists of telling 'Just So' stories and finding evidence to fit them. But the approach eventually yielded rich rewards after tightening up its act. As leading behavioural ecologists Martin Daly and Margo Wilson showed in their study of child abuse and step-parenting, the evolutionary approach provided, for example, a much greater insight into the nature of relationships between relatives and non-relatives.

A decade or so later came the evolutionary psychologists – the psychological rediscovery of Darwin. History repeated itself with more sloppy thinking, more controversy – and rewards? Well, some interesting conjecture about our own evolution.

Evolutionary psychology is controversial because testing hypotheses about human behaviour is confused by the influence of culture. Some problems might be avoided by studying preliterate societies, but there are few of them left. The alternative, used by many evolutionary psychologists, is to search for what they call 'universals' – traits that are perceived in the same way across all human cultures. This is what David Buss has done. What he found, by studying a number of cultures, was that what men find attractive in women is youth and beauty because all the features that make women beautiful, such as clear skin, luxuriant hair, an hourglass figure and so on, are signs of fertility. In the cold hard light of evolution, what males want (usually unconsciously) are fertilisations and descendants. Women are programmed to do exactly the same, but they achieve it in a different way.

What are women choosing? The answer, say evolutionary psychologists such as Buss and Robert Wright, author of *The Moral Animal*, is that women choose men less on the basis of looks and more on status, resources and a willingness to share those resources – particularly in their choice of long-term partners. Women need resources in order to rear their babies, and male resources usually come in the same package as status. In preliterate societies such as the Yanomami Indians from South America, head men have more wives, more extra-marital affairs and more offspring than other men. Of course, the association between high male status and reproductive success may arise partly because of male–male competition rather than female choice, and indeed in circumstances like this it is often difficult to distinguish between the two.

Social scientists are critical of the evolutionary psychologists' studies because they focus on the differences in what each sex wants from a partner and ignore the broad similarities, which they consider to be much greater. Evolutionary psychologists counter this with the argument that concentrating on the similarities and ignoring the differences is like thinking chimpanzees and bonobos are really human because they share 98 per cent of their DNA with us.

However, social scientists do acknowledge that women are attracted to men with status. So how do men acquire status? The answer is: in any way they can. When we think of high status in humans we tend to think of presidents and millionaire football players, but status is relative and men compete for it in a multitude of ways. In *The Mating Mind*, Geoffrey Miller suggests that the massive and rapid evolution of brain size in humans occurred as a direct result of sexual selection – females preferentially choosing successful men.

Miller focuses on the spectacular: Picasso's prodigious output, Balzac's *Comédie Humaine* and Rachmaninov's Third Piano Concerto, but even a modest display is better than none at all. In poor, rural parts of Spain, running with the bulls is a great prestige-enhancing strategy – it is risky, but the pay-offs are considerable. If you are less young, or less fit, then there is always the pigeon contest in which men pit their male pigeons against others to see which one a single female pigeon chooses. Not especially risky, nor very sexy, but better than nothing. In Britain, the equivalents are being a member of the local football team or, if you are somewhat less fit, the darts team.

Overall, men compete more often and go to greater extremes than women in everything. Just look at *The Guinness Book of Records* under the miscellaneous human achievement category. Men vastly outnumber women in their miscellaneous but extreme activities. What are now needed are studies designed to test evolutionary psychologists' ideas that all male human endeavour is motivated by sex, whether they realise it or not. For example, one could measure male reproductive success and see whether it correlates with male status, within peer groups rather than on a global scale. The problem is, it would be necessary to use molecular paternity analyses to catch any extra-pair offspring, and for ethical reasons this would be extremely difficult to carry out.

The classical view of sexual selection, propounded by, for example, Andersson in *Sexual Selection*, is that there are a few high-quality men out there that all women should be keen to mate with. If women do manage to mate with them, what do they gain? The first possibility is that more attractive males are genetically superior, and there is much debate about whether this is true even in non-humans. The other, less contentious, possibility is that more attractive males have more resources that females can use to help rear their offspring. While male status and the resources it brings do seem to be universally admired by women, there are also much more subtle things going on. Mate choice is complex, and both humans and non-humans use a wide range of information in choosing partners, some of which is gleaned from experiences during development.

If, for example, young zebra finches are reared by Bengalese finch foster-parents, on reaching sexual maturity these zebra finches prefer Bengalese finches to individuals of their own species as sexual partners – a phenomenon known as sexual imprinting. In *Keeping the*

Love you Find, Harville Hendrix suggests that something similar occurs in humans. He says the key point is to ask why we have evolved in such a way that our choice of partner is influenced by our parents' behaviour. Why is it adaptive for men to choose a partner who resembles their mother and for women to choose someone who resembles their father? One possible answer, originally proposed by Patrick Bateson in Cambridge on the basis of a study of quail, is that, providing we avoid breeding with close relatives, individuals that breed with individuals who are genetically and culturally similar to themselves leave more descendants than those who do not.

Recent studies also suggest that body odour may say something about a male's quality – not in an absolute sense, but relative to that of a female. The major histocompatibility complex (MHC) is a set of genes responsible for the ability to combat infection. Molecular techniques now allow us to 'type' an individual's MHC and it turns out that this set of genes varies enormously between individuals. Male mice advertise their MHC type through an odour in their urine. Given a choice, females choose to mate with males whose MHC differs from, that is, compliments, their own. A female mouse living in the territory of a male whose MHC is the same as her own will attempt to seek extra-pair matings with a male of another MHC type. Much the same happens in people, in that when given a choice of male body odours, women prefer those of men whose MHC differs from their own. This research, conducted by Claus Wedekind at the University of Edinburgh, is controversial, but his findings make evolutionary sense. Spontaneous abortion (of very early embryos) is much more common in humans than was previously thought, and is especially common among couples whose MHCs are similar. So there may be genetic advantages in choosing particular males after all.

Tim Birkhead
Professor of behavioural ecology at the University of Sheffield

What makes us fall in and out of love?

David M. Buss
Professor of psychology at the University of Texas

? Why someone falls in love with one person rather than another, from the thousands of potential partners they encounter, remains a mystery of profound proportions. Chance, chemistry and the odds that two small windows of receptivity will open simultaneously at the precise moment of meeting guarantee unpredictability. Despite this mystique, science has made modest inroads into understanding the ins and outs of love.

Contrary to common myths disseminated in the social sciences in the twentieth century, love was not invented by Western European poets a few centuries ago. Evidence points to the opposite conclusion – love is a cross-cultural universal and probably has been since the emergence of long-term pair bonds back in the mists of human evolutionary history. From the Zulus in South Africa to the Inuits in north Alaska, people report experiencing the obsessions of mind and passions of emotion that those in the Western world link with love.

In a survey of 168 diverse cultures, anthropologist Bill Jankowiak found strong evidence for the presence of romantic love in nearly 90 per cent. For the remaining 10 per cent, the anthropological evidence was too sketchy for definitive conclusions.

Many people worldwide also report being currently in love. Sociologist Sue Sprecher and her co-authors interviewed 1,667 women and men in Russia, Japan and the United States. They found that 61 per cent of the Russian men and 73 per cent of the Russian women reported being currently in love. Comparable figures for the Japanese were 41 per cent of the men and 63 per cent of the women;

among Americans, 53 per cent of the men and 63 per cent of the women.

My study of the mate preferences of 10,047 individuals from thirty-seven different cultures located on six continents and five islands also revealed the importance and universality of love. I found that 'love and mutual attraction' was rated as the most indispensable of eighteen possible attributes of someone they might marry – by both sexes, in all cultures. Through the vagaries of cultural prescriptions, diversity of mating systems, roils of political regimes, disparities of economic conditions and multiplicity of religious exhortations, humans everywhere apparently long for love.

The essential qualities people desire in a mate define the ground rules of human mating. Desires determine whom we are attracted to and which tactics are effective in attracting them. Violations of desire create conflict and predict conjugal dissolution. Satisfying another's desire becomes an effective means to gain and retain a mate. Fulfilment of desire increases the odds of long-term love.

The thirty-seven cultures study illuminated more precisely than before what these components of desire are. People the world over want mates who are kind, understanding, intelligent, dependable, emotionally stable, easy-going, attractive and healthy. But cultures differ tremendously on the importance they place on some qualities. Virginity, for example, was judged to be virtually indispensable in a potential mate by most mainland Chinese, but irrelevant by most Swedes and Dutch.

What was surprising to social scientists was the discovery of universal sex differences. Men worldwide place more importance on youth and physical attractiveness, qualities now known to be important signals of a woman's fertility and future reproductive potential. Women across the globe want men who are ambitious, enjoy decent social status, possess resources or the potential to acquire them, and who were born a few years before they were. Over the long expanse of human evolutionary history, a woman's children survived and thrived better by selecting a resourceful man who committed to her.

But is love a cold-blooded appraisal of a person's spec sheet or an emotion that blinds us to deficits? It is a bit of both. People rarely fall in love with those who lack the qualities that they desire. In a study of women's and men's responses to personal ads, men were more likely to initiate contact with women who mentioned physical attractiveness and a young age in their self-descriptions. Women were more likely to initiate contact with men who mentioned a decent income and a respectable education level.

While who we love often has a ruthlessly utilitarian logic, love may have evolved to blind us to a partner's deficiencies. There are at least two scientific explanations for love's myopic effects. Few people possess the full complement of desired qualities and most of us have to settle for less than we would want in an ideal world. Usually, only those high in desirability can attract others comparably high. Perhaps the most scientifically documented law of love is assortative mating, the pervasive coupling of people who are similar to each other. Intelligent, educated people marry those who share their insights and erudition. The glamorous pair off with the glamorous. Although opposites occasionally attract, when it comes to long-term love the '8's' typically marry the other '8's', while the '6's' go with the other '6's'.

It would not pay to harp on deficiencies while falling in love. In fact, a recent study reported that most people show 'love's illusion' of being overly optimistic about their chances of marital success. Whereas roughly 50 per cent of marriages will end in divorce, only 11 per cent of married people thought that their own marriage would end in divorce. Among a younger group of unmarried individuals, only 12 per cent thought that their future marriage would have a 50 per cent chance of splitting. Among those marrying now, the likely divorce rate has increased to 64 per cent. These findings may reflect adaptive biases that, although clearly off target, do function to increase the odds of success. According to this explanation, love is an emotion that motivates people to persevere through thick and thin, even if it does not always work out in the end. In short, love can blind us in two ways – first, by making us happy to settle for someone who is less than our fantasised ideal, and second, by being optimistic about the future of the romance, and thereby enhancing its chances.

Evolutionary economist Robert Frank argues that love is a solution to the problem of commitment. If a partner chooses you for rational reasons, he or she might leave you for the same rational reasons, finding someone slightly more desirable on all of the 'rational' criteria. This creates a commitment problem: how can you be sure that a person will stick with you? If your partner is blinded by an uncontrollable love that cannot be helped and cannot be chosen, a love for only you and no other, then commitment will not waver. Love overrides rationality. It is the emotion that ensures that you will not leave when someone more desirable comes along and provides a signal to the partner of long-term intent and resolve.

It is likely that the causal arrow also runs in reverse. Love may be the psychological reward we experience when the problem of commitment is successfully being solved. It is a mind/body opium that signals that the adaptive problems of mate selection, sexual congress, devotion and loyalty have met with triumph. The scientific explanation is that evolution has installed in the human brain reward mechanisms that keep us performing activities that lead to successful reproduction. The downside is that the drug wears off. Some get on a hedonic treadmill, chasing the high that accompanies love. Repeating successful mating with fresh partners brings back the buzz, but perhaps never to its former level.

Love may be a solution to the commitment problem or an intoxicating reward for successfully solving it, or both. But there is no question that love is an emotion intimately linked with commitment. In my studies of 115 different actions that signal whether a person is truly in love, acts of commitment topped the list, such as talking about marriage or expressing a desire to raise a family. The most

Big Questions in Science

salient acts of love signal the commitment of sexual, economic, emotional and genetic resources to one person.

Unfortunately, that is not the end of the evolutionary story. Once desire for love exists, it can be manipulated. Men deceive women about the depth of their loving feelings to gain short-term sexual access. Women, in turn, have co-evolved defences against being sexually exploited, for example by imposing a longer courtship process prior to consenting to sex, attempting to detect deception and evolving superior ability to decode non-verbal signals. The co-evolutionary arms race of deception and detection of deception continues with no end in sight.

Another problem is that people fall out of love as crushingly as they fall in it. We cannot predict with certainty who will fall out of love, but recent studies provide some clues. Just as the fulfilment of desire looms large when falling in love, violations of desire portend conflict. A man chosen in part for his kindness and drive may get dumped when he turns cruel and lazy. A woman chosen in part for her youth and beauty may lose out when a newer model beckons her partner. An initially considerate partner may turn condescending. And a couple's infertility after repeated episodes of sex may prompt each to seek a more fruitful union elsewhere.

Then we must consider the harsh metric of the mating market. Consider an entry-level professional couple. If the woman's career skyrockets and the man gets fired, it puts a strain on both because their market values now differ. To the woman, a '9' who was previously out of reach now becomes available. In the evolutionary jungle of mating, we may admire a woman who stands by her loser husband. But those who did are not our ancestors. Modern humans descended from those who traded up when the increment was sufficient to outweigh the manifold costs people experience as a consequence of breaking up.

Falling out of love has many dark sides. The crash can be physically dangerous for women and psychologically traumatic for both sexes. Men who get rejected by the woman with whom they are in love often abuse them emotionally and sometimes physically. In our recent studies, we found that an alarming number of men who are unceremoniously dumped begin to have homicidal fantasies. Just as evolution has installed reward mechanisms that flood us with pleasure when we successfully mate, it may have also equipped us with mechanisms that deliver psychological pain when we experience mating failure.

Further reading

Malte Andersson: *Sexual Selection* (Princeton University Press, 1994)

Tim Birkhead: *Promiscuity* (Faber & Faber, 2000)

David M. Buss: *The Evolution of Desire: Strategies of Human Mating* (Basic Books, 1995), *The Dangerous Passion: Why Jealousy is as Necessary as Love and Sex* (Bloomsbury, 2000)

Martin Daly and Margo Wilson: *The Truth about Cinderella: A Darwinian view of parental love* (Weidenfeld & Nicolson, 1998)

Stephen Jay Gould and Richard C. Lewontin: 'The Spandrels of San Marcos and the Panglossian paradigm: A critique of the Adaptationist Programme' (*Proceedings of the Royal Society of London B, Biological Sciences* 205:581–98, 1979)

Geoffrey Miller: *The Mating Mind* (Heinemann, 2000)

Robert Wright: *The Moral Animal: Evolutionary Psychology and Everyday Life* (Vintage, 1994).

What Causes Aggression ?

A solitary matador and a half-crazed bull replayed their parts in an óld, violent ritual in a sun-soaked arena in southern Spain. It was the summer of 1964, but film of the incident is still shown in lecture halls today. As the bull bore down on the unarmed man, it became apparent that the red cape was not being wielded with a matador's usual poise. It was limp and motionless. In fact, the man at the centre of the ring, the brain scientist José Delgado, had never faced a charging animal in his life.

But the horns never reached the doctor. Seconds before impact, Delgado flicked a switch on a small radio transmitter he was holding and the bull immediately braked to a halt. He pressed another button and it meekly turned to its right and trotted away.

Delgado was triumphant. After fifteen years studying the workings of the brain, he had proved in the most dramatic fashion that understanding and controlling its mechanisms had reached a refinement that allowed an animal's aggression to be turned on and off by remote control. He explained that he had been playing monkeys and cats 'like little electronic toys', making them fight, mate and go to sleep using the same technique of inserting probes in the brain and electrically stimulating relevant tissues.

'A turning point has been reached in the study of the mind,' he announced. 'I do believe that an understanding of the biological bases of social and antisocial behaviours and of mental activities, which for the first time in history can now be explored in a conscious brain, may be of decisive importance in the search for intelligent solutions to some of our present anxieties.' It was the high point of scientists' belief in their ability not only to explain but to intervene in the workings of the brain and, specifically, in their ability to prevent antisocial aggression.

Today, Delgado and his contemporaries are still regarded as an inspiration by many of their successors. Felicity Huntingford, an expert on animal aggression at Glasgow University, describes Delgado as 'a pioneer and a brilliant science communicator who showed us what could be done. He started a lot of people on this path.'

But there was a darker side to the mid-twentieth century's boundless confidence in the potential for physiological explanations and treatments for brain processes, according to Professor Huntingford. In the thirty years preceding Delgado's showmanship, an estimated 40,000 to 50,000 people were lobotomised in the United States alone, often to prevent abnormal aggression.

Walter Freeman, the relentless populariser of the treatment in the US, was confident that pushing a medical 'ice pick' into patients' brains and destroying tissue near the 'thalamus', an area of the brain that he theorised was responsible for emotional overexcitement, would help remove the causes of aggression and a range of other problems. There was little proof for his theories, but thousands of people were turned into vegetables by the operations (including, most famously, John F. Kennedy's sister, Rosemary).

Controversy has since raged about more academically respectable treatments, such as that proposed by the neurosurgeon Vernon Mark and psychiatrist Frank Ervin in 1970. In *Violence and the Brain*, they claimed, like Delgado, to be able to determine with precision parts of the brain responsible for aggression. They claimed great improvements in behaviour once the 'problem areas' had been destroyed using electrodes, although others thought serious brain damage was the main result. Chemical methods of reaching the same parts of the brain, such as the so-called chemical lobotomy drug Thorazine, which is still popular, have provoked similar passionate disagreement.

Why do lobotomies, chemical lobotomies and similar treatments prompt such disquiet? They have often been highly effective at treating aggression. Whatever the imperfections of Freeman's understanding of the brain, he was doing something right – regularly turning uncontrollably aggressive people into unusually docile and compliant members of society. Yet there was something profoundly disturbing about the smiling zombies that he and others were often responsible for creating.

'What people did not, and some still do not, seem to understand is that our capacity for aggression and other mental processes are tremendously complex and intertwined with each other,' says Professor Huntingford. 'It is not a question of just removing this bit of a brain and fixing a problem. You are dealing with something that is very close to the centre of what makes us what we are.'

Anthony Burgess made the point in his novel *A Clockwork Orange*. This gives a shocking account of the way the extreme aggression of the book's anti-hero, Alex, is successfully treated using a brutal form of aversion therapy. By the end of his course Alex is no longer a threat to society, but he has also been deprived of a crucial aspect of his humanity. As Burgess saw it, man is fundamentally both 'gloriously creative and bestially destructive'. It is impossible to have one without the other.

Understanding of the physiology of the brain has come a long way since Freeman's simplistic phrenology. Recent research has focused on neurotransmitters – chemicals that transmit messages between brain cells – and specifically on a transmitter called serotonin that appears to be strongly related to aggressive behaviour. Studies have found that giving animals drugs that lower serotonin levels sometimes makes them more aggressive and that increasing levels can have the opposite effect.

But the role of serotonin is not straightforward. The brain has at least fourteen serotonin receptors and researchers do not yet know exactly the role of each. Levels of the neurotransmitter are also associated with problems such as depression and eating disorders, and the behaviour of some of the receptors is highly complex. For instance, one receptor, known as 1B, appears to decrease aggression in mice and monkeys when activated. However, the effect is associated with a lowering in serotonin levels rather than the expected increase.

Another series of studies associating the neurotransmitter vasopressin with aggression has further muddied the waters, indicating that aggressive behaviour may be governed by a complex interplay of chemicals in the brain.

Genetic research has revealed a similarly confusing picture. There have been dramatic results. One study, for example, traced a history of extreme violence in a Dutch family back to defects in a gene governing the breakdown of neurotransmitters. But strident headlines claiming that researchers are about to identify aggression genes ignore difficulties in interpreting these results. A lab experiment on mice, for example, could show an increase in fighting because a gene that governs their sense of smell has been disabled – the animals could be unable to use pheromones to signal to each other and therefore avoid conflict.

Another set of researchers insists that such mice studies are of extremely limited use in understanding human violence anyway. Adrian Raine at the University of Southern California believes our highly developed prefrontal cortex, which is tiny in lower mammals such as mice, has a vital role in governing aggression. Research has shown that murderers have unusually low levels of glucose metabolism in their prefrontal cortex, indicating problems in that part of the brain.

It is an illustration of the complexity of the subject, however, that most theories of aggression do not even look at the physical make-up of the brain. Approaches since the 1920s have ranged from Sigmund Freud's theory that aggression is an externalisation of every human's death wish and exists with or without external provocation, to Konrad Lorenz's conclusion, based on ethological studies of fish and birds, that it is a drive shared by man with most of the rest of the animal kingdom.

One of the most fertile approaches, contradicting the tendency of biological theories to concentrate on the brain state of individuals, has looked at the crucial role of social forces in motivating and governing aggressive behaviour. Consider the causes of the terrorist attacks on the World Trade Center and Pentagon in September 2001. While some of the suicide bombers might conceivably have had low levels of serotonin in their systems, religion, culture, ideology, the influence of authority and history all seem to be more directly relevant in explaining why they killed so many people.

Humanity's capacity for intellectual thought, learning and complex communication has often been used to elevate us above animals, but these qualities may also contribute to aggressive behaviour. Studies have shown that children who are shown violence are much more aggressive when provoked than those who have been shown pacific behaviour. A review in 1994 of thirty-five years of research into the relationship between Western television's violence and aggression found a significant positive correlation. Meanwhile, traditional societies on Tahiti and among the Inuit are reported by anthropologists to show little antisocial aggression because of strong cultural disapproval of aggressive behaviour.

Leading social learning researcher Albert Bandura even questions the relevance of the concept of aggression to understanding much of man's inhumanity to man: 'What I am really interested in is not aggressive feelings but moral standards and how very skilled we are at disengagement from these. That seems to be the real determinant of whether we will harm someone: whether we have removed them from moral consideration.'

To understand why a man would fly a passenger plane into a

building containing tens of thousands of civilians, or even why someone would steal an old lady's purse, it is perhaps less important to know how aggressive he was feeling that day than to understand what he thought would be achieved by the act and how he had convinced himself to ignore his victims' suffering.

Chris Bunting
Freelance journalist

What causes aggression?

Dolf Zillmann

*Burnum Distinguished Professor of Information Sciences
and Psychology at the University of Alabama*

Analysis of aggression among animals reveals a basic set of conditions under which more or less all species, especially those towards the upper end of the evolutionary scale, resort to inflicting injury and death upon other species, as well as upon members of their own. Predatory behaviour for sustenance is the most obvious cause of destructive behaviour against another species. Procuring territories that provide the conditions needed for subsistence instigates further violence. Roughly speaking, these situations translate to aggression for food and shelter, perpetrated in the interest of self-preservation. Within-species aggression serves the same purpose in situations where resources are scarce.

Additionally, within-species aggression pursues reproductive ends, as males fight for sexual access to receptive females. Such aggression might be considered to serve the preservation of the species. It is generally held, though, that in contrast to predatory aggression, within-species fighting does not aim at inflicting injury and death. Its object is to induce rivals to yield in the competition over resources.

Is aggression among humans precipitated by the same circumstances? Some scholars suggest that aggression-provoking conditions are essentially the same for all species and that there is nothing unique about human aggression. After all, humans liberally slaughter other species for nourishment, fight wars with each other over territories, inflict injury and death upon others to usurp their valuables, and resort to violence to defend whatever they hold dear. Quite often, sexual rivalries also escalate to brutalities. Humans, just like many other species, thus seem prepared to use destructive force to get what they want.

Other scholars find these analogies wanting. They point to the evolution of the human neocortex and the fact that its mental capacity far exceeds that of other species. Moral contemplation – the ability to judge what is good and right under given circumstances – and volitional control – in this case the ability to bring one's actions in line with moral assessments – take centre stage in such theorising. These scholars may concede the existence of archaic aggressive impulses, but believe that, as a rule, rationality is capable of squelching them. They therefore contend that little, if anything, can be learned from analyses across the human/non-human divide.

I take a more integrative stance, acknowledging both our ancient evolutionary heritage and the comparatively recent expansion of our cognitive faculties.

It is generally held that evolution of the brain progressed from a reptilian core, which was encapsulated by palaeomammalian structures, known as the limbic system, which, in turn, were encapsulated by neomammalian structures, the neocortex. Notwithstanding development of an especially large neocortex, our brains have retained the tripartite structure that integrated the earlier evolved, ancient structures. More importantly, these structures continue to exert their influence on all vital human behaviours much as they did for millennia.

The limbic system controls all human emotions, and the amygdala, a part of this system, has emerged as the most significant structure in controlling aggression. This structure is involved in monitoring the environment for cues of danger and, when it encounters them, in initiating endocrine processes to help people cope effectively with them in a physical way.

Coping with immediate threats of danger requires an individual to be provided instantly with energy for vigorous action, primarily for warding off a threat by attacking it or for eluding it by retreating swiftly. The energy needed is mediated by systemic release of mostly adrenal hormones that foster sympathetic excitement, in large measure by providing glucose to the skeletal muscles.

This set of responses defines the well-established fight/flight reaction, which is ideal for behavioural emergencies that can be resolved by an episode of forceful action. In evolutionary terms, the mechanism for such action has served the species well. It has helped humans to survive unexpected confrontations with predators and hostile fellow humans. Becoming roused and agitated, feeling strong and able to rise to a challenge, and focusing entirely on the here and now in coping with danger proved to be highly adaptive.

It is only in modern society that this adaptive value has been compromised. As a rule, threats of danger can no longer be countered with direct assault or spontaneous escape. The adverse consequences of radon gas in one's home, for example, cannot be eliminated by instant physical action, no matter how much rousing energy the body provides for coping with the apparent emergency. Neither does it help us to cope by fight or flight with problems such as taxation or global warming. Probably most important is that societal sanctions curtail, by threat of punishment, resolution of common conflict through violent or evasive action. It is inadvisable for someone whose car gets dented by a careless driver to beat him up in a fit of rage, and it is usually not feasible for someone who owes child support to flee the country on impulse. All such cases of provocation and frustration nonetheless engage the archaic brain structures to initiate potent reactions, in spite of the fact that these reactions have largely lost their utility. This often fosters unbearable anger and triggers violent action that is incapable of removing the grounds for the emotion.

In understanding the emotions of fear and anger it is important to recognise their initial function as well as their more recent dysfunction. The initial function was twofold. It was to furnish energy for a bout of action and to focus attention on the here and now of the action. These two responses, known as action impulsivity and cognitive deficit, still characterise anger and rage. One urges aggressive action irrespective of that action's ultimate usefulness. The other, because of cognitive preoccupation with the immediate situation, makes individuals negligent of all non-immediate implications of their actions. This deterioration of cognitive control, which makes people oblivious to the consequences of their violent actions, is obtrusive enough to be widely considered a form of temporary insanity, mitigating responsibility.

The propensity to commit acts of destructive violence undoubtedly resides in all of us. Threats of harm and degradation elicit reactions that, when reaching extreme levels, are bound to lead to uncontrolled, impulsive, aggressive behaviour. Residues from unrelated frustrations and challenges of everyday life often enter into the reaction to specific circumstances. Because anger can be fuelled by stimulation from different sources, seemingly minor disagreements often escalate into fury and violent conflict.

So far, we have considered the influence – at times dysfunctional – of archaic brain structures. Now we shall contemplate the influence of the new structures that set us apart from other species: the neocortex with its associative, anticipatory and inferential powers.

Most scholars of aggression embrace the view that superior rationality, afforded us by the neocortex, is the antidote to violence. Rationality is seen as the panacea to all baser human compulsions. No doubt rationality can, and often does, prevent violent outbursts. But even cursory inspection of the records of impulsive, destructive violence shows that, when it comes to preventing violence, reason regularly fails us.

More importantly, rationality not only fails to provide an effective antidote to violence, it is the direct cause of enormous amounts of violence perpetrated by humans against each other. It is our ability to reason that tells us that taking others' valuables by force in a way that manages to minimise or entirely bypass repercussions is a winning formula. Our anticipatory skills are accordingly put to use in constructing strategies that make violence pay. They not only place every individual at risk of coercing others by way of aggressive action, they also inspire organised violence and warfare.

Human aggression is not merely helped by superior anticipatory skills and the strategies they yield. It is also driven by what some consider the very highest form of rationality: moral reasoning. The moral concepts of equity and retribution constitute major sources of aggression. Social justice comparisons that place us at the short end of rewards, despite comparable efforts, are infuriating and thus instigate aggression. Violation of our sense of justice demands retaliation. If we have been wronged, we must 'get even'. The desire to retaliate in order to set things right leads frequently to interpersonal clashes. Wars are often fought because someone persuaded the populace that past humiliations could not go unpunished. On occasion, even the perpetration of the vilest of atrocities is construed as morally mandated, usually by reference to divine authority.

So the neocortex, while enabling us to recognise the social ills and global dangers of violence, has furnished new and uniquely human reasons and avenues for aggression. Both the thoroughly reasoned conception of strategies for effective aggression and the moral mandate for aggression are not to be found in other species. These motives for aggression, then, set us apart from other animals. At the same time, however, we continue to share with other primates, and other species, the motives for aggression that reside in the archaic structures of our triune brains.

Further reading

John Archer: *The Behavioural Biology of Aggression* (Cambridge University Press, 1988)
Albert Bandura: *Aggression: A social learning analysis* (Prentice-Hall, 1973)
Ernest H. Johnson: *The Deadly Emotions: The role of anger, hostility and aggression in health and emotional well-being* (Praeger, 1990)

Roger N. Johnson: *Aggression in Man and Animals* (Saunders, 1972)

Arnold S. Linsky, Ronet Bachman and Murray A. Strauss: *Stress, Culture and Aggression* (Yale University Press, 1995)

Ashley Montagu: *The Nature of Human Aggression* (Oxford University Press, 1976)

John Renfrew: *Aggression and its Causes: A Biopsychosocial Approach* (Oxford University Press, 1997)

Dolf Zillmann: *Connections between Sexuality and Aggression* (Erlbaum, Lawrence Associates, 1998).

Big Questions in Science

Is It Right to Interfere With Nature ?

In July 2001, the American House of Representatives rejected a motion to allow human cloning, together with an amendment to allow the more limited therapeutic cloning. Speaking in the debate, the Republican congressman for Oklahoma, J.C. Watts, declared: 'This House should not be giving the green light to mad scientists to tinker with the gift of life. Cloning is an insult to humanity. It is science gone crazy.' Three months later, two such 'mad scientists' – pioneers of stem cell techniques and in vitro fertilisation – were honoured with a Lasker Award, America's equivalent of a Nobel. In the scientific community, their work was seen as a valuable contribution to increasing human knowledge.

Again and again, human curiosity and the scientific developments that result from it come up against human caution about possible side effects and fears that the hidden consequences of a discovery will outweigh the benefits. The American debate on cloning tapped into widespread public suspicion of scientists and the belief that scientific tampering with nature must be wrong.

This fear is not new. But for Lewis Wolpert, professor of biology as applied to medicine at University College London, it is misguided. He has promised a bottle of champagne to anyone who can give him one new ethical issue that human cloning raises, and draws a distinction between science and its application – technology. He argues that science itself is value-free; it is what society chooses to do with it that has a moral value. In *The Unnatural Nature of Science*, he discusses the development of the atomic bomb that killed 200,000 people in Hiroshima in 1945. The bomb exploited Einstein's equation '$E=mc^2$', published in 1905 in his *Special Theory of Relativity*, which simply states that a large amount of energy can be released from a small amount of material. The decision to build the bomb, says Wolpert, was political rather than scientific. His argument is that it is impossible to block the advance of knowledge. While scientists must inform the public of possible implications of their work and its reliability, the application of knowledge is not their responsibility.

Einstein himself would have disagreed. Together with philosopher Bertrand Russell, in 1955 he launched a manifesto calling upon scientists of all political persuasions to discuss the threat posed by thermonuclear weapons. The first meeting, in 1957, of what became known as the Pugwash Conferences attracted twenty-two eminent scientists, and the conferences continue to meet regularly today.

Big Questions in Science

The view that knowing too much about nature is a bad thing is played out in myths central to the Western culture from which modern science has evolved. Ancient Greek myths tell how, in return for stealing fire from the gods and giving it to mankind, the Titan Prometheus is chained to a rock while an eagle pecks out his liver, which regrows daily. Adam and Eve are banished from Paradise when they eat from the tree of knowledge. In John Milton's version of the story, *Paradise Lost*, written in 1667, the serpent addresses the tree as 'Mother of Science'. Playwright Christopher Marlowe's medieval alchemist Dr Faustus was prepared to sell his body and soul to the Devil in exchange for knowledge.

Mary Shelley's gothic novel of 1818, *Frankenstein, or the Modern Prometheus*, has become the model for scientists who want to know too much and try to interfere with nature. Victor Frankenstein, a young idealistic student, discovers the secret of life. He creates a living being, which although initially gentle and loving, becomes embittered by the fear and rejection he receives from society, wreaking a hideous revenge on his creator.

Over the past five years, the 'Frankenstein' prefix has become almost synonymous with 'genetically modified', the practice of altering the genome of an organism by removing or inserting genes, sometimes from other species. Tabloid newspapers coined the term when vegetarian cheese and tomato paste containing ingredients that had been modified were introduced to the sceptical UK market. Public outrage soon led to retailers withdrawing such products and declaring themselves GM-free. Meanwhile, newspapers talked of Frankenstein forests, Frankenstein fish and Frankenstein babies. The implication was that scientists had delved too deep into nature and that the consequences could only be dire.

Popular opinion often tends to the view that interfering with nature can only lead to bad outcomes. Scientists' claims that genetically modified crops could help feed developing countries and reduce amounts of chemical fertilisers and pesticides were largely ignored by popular opinion in Britain in the late 1990s. Protesters broke into fields and destroyed field trials of GM crops, claiming that the technology was unproven and uncontrollable. In 1999, Prince Charles called for a public boycott of GM foods, claiming that they were unnecessary and environmentally dangerous. Prime Minister Tony Blair issued a counter-warning of the dangers of becoming anti-science, adding that it was wrong to make heroes of people preventing basic scientific research.

Fears about new scientific discoveries do have their foundation. For example, scientists deemed the sedative thalidomide so safe that it was prescribed for pregnant women to control morning sickness. The first thalidomide baby was born in 1956 and the drug was taken off the market in 1961 after it was linked to severe birth defects, high rates of miscarriage and an estimated 40 per cent mortality rate before the victims' first birthdays.

But other technologies initially scorned and rejected by popular opinion have proved to be beneficial and become widely accepted by the mainstream. In 1798, Edward Jenner discovered that inoculating people with a small preparation of cowpox (*vaccinae*, from the Latin word for cow, *vacca*) would make them immune to smallpox. This 'vaccination', as it came to be called, saved many millions of lives and, in 1980, led to the World Health Organisation declaring the disease eradicated. However, at the time of its discovery, political cartoonists published engravings showing people who had been vaccinated sprouting cows' heads. Some were afraid of being injected with material from cows, saying they would not be treated with substances from God's lowlier creatures. When vaccination with cowpox was made compulsory in 1853, there were protest marches.

Public jitters also followed the birth of Louise Brown, the first test-tube baby, in 1978. The pioneering work of Robert Edwards and Patrick Steptoe, the British researchers who perfected in vitro fertilisation of the human egg, prompted concerns about designer babies, eugenics and human/animal hybrids. This led the government to create a committee of inquiry in 1982 under Dame Mary Warnock. The Warnock Report, published in 1984, recommended setting up a regulatory body to govern all fertility treatment. The Human Fertilisation and Embryology Authority (HFEA) effectively enacted its proposals in 1991.

All embryos handled in the UK must be licensed by the HFEA, whether they are for storage or for use in fertility treatments or research. Since its inception, it has issued licences for the creation of almost one million embryos, about 6 per cent of which were donated to research. Licences are granted for research into miscarriage, infertility treatments and contraception. Changes to the legislation in early 2001 have allowed embryos also to be used in research into embryo development and to increase knowledge of serious disease. It means that the HFEA would also be responsible for research into human cloning.

This became an issue debated by governments around the world after cloning techniques were showcased by a team at the Roslin Institute in Edinburgh in 1997. The Roslin team introduced Dolly the sheep, which had been created by somatic cell nuclear transfer. In this, the nucleus of a non-reproductive cell is transplanted into an egg with the nucleus removed. If the egg is fertilised, the resulting offspring will carry the genetic material of the transplanted nucleus rather than that of the original egg. Stem cells are present in embryos for a limited period in their early development. These cells have the potential to develop into any tissue type in the body. In the future, scientists hope to combine the two techniques, creating cloned embryos of patients, harvesting the stem cells and using them to grow matching tissue to replace damaged organs. Scientists believe that this process, known as therapeutic cloning, offers hope in the long-term to patients suffering from degenerative conditions including heart disease, Parkinson's disease and Alzheimer's.

While the House of Representatives voted against human cloning in July 2001, the UK's House of Lords legislated to allow therapeutic cloning from January 2001. The HFEA may now grant licences to researchers to create embryos from which to derive stem cells, although scientists are not allowed to implant an embryo into a woman to allow it to grow into a person.

But already an Italian embryologist wants to go further, rousing international indignation by saying he plans to apply the techniques used to create Dolly to humans. Severino Antinori says he has begun trials to clone human beings for infertile couples, and claims two hundred couples have volunteered to take part. Giuseppe Del Barone, head of Italy's medical council, has threatened to banish him from Italy if he carries out the trials, describing the work as 'the rape of nature which goes against human dignity'. A German newspaper has dubbed Antinori 'the Italian Frankenstein'.

Concerns about scientific interference with nature are now focusing on the Human Genome Project and publication of the first draft of the 'Book of Life'. Its publication met a fairly positive response, but a number of worries have been raised about how the data will be applied. Thorny moral issues already being debated include the possibility of prenatal diagnosis leading to increased rates of abortion and the use of genetic testing by insurance companies to minimise risks. Five per cent of the project's annual budget was committed to studying ethical, legal and social implications.

But many scientists agree with Wolpert that ethical issues can be a distraction from the science. They believe that the US clampdown on stem cell research could precipitate a reverse brain drain, with American scientists moving to Britain, attracted by greater freedom to practise. One prominent researcher, Roger Pedersen, quit his laboratory in California for Cambridge to carry out publicly funded research on stem cells, citing the difficult political climate in the United States as one reason for leaving. Some American biotech companies have also decided to invest in cloning companies outside the United States.

Ultimately, human 'nature' has never been able to resist the urge to increase human knowledge. But concern over the dangers of interfering with nature is just as long-standing and seems equally 'natural'. The struggle between the two looks set to run and run until we know everything – or have destroyed ourselves in the attempt to find out.

Caroline Davis
Reporter on the Times Higher Education Supplement

Is it right to interfere with nature?

Mary Warnock
*Moral philosopher, former mistress of Girton College,
Cambridge, and chair of the Committee of Inquiry
into Human Fertilisation and Embryology 1982–1984*

? The word interference is, on the whole, pejorative: those who argue that the new biotechnology and its possible applications constitute an interference with nature certainly understand it to be so. But when Prince Charles, in his 2000 Reith Lecture, besought biologists to learn more about nature, if they wanted, but not to seek to change it, his father and sister were quick to point out that men had been changing nature – interfering with it – since human history began. And it is true that, had they not interfered, we would be living a wilderness life, none of our sophisticated pleasures and pursuits even conceivable. Although this is well understood (and certainly, in fact, by Prince Charles, who was really pleading not for non-intervention, but for the use of 'tried and tested' techniques of agriculture), there is a persistent appeal in the cry that biological science has now gone too far, and that we ought to stop interfering and let nature take its course.

Our idea of what counts as natural is complex and has changed through the centuries. At the present time, it has at least two crucial components, the romantic and the Darwinian.

French philosopher Jean-Jacques Rousseau began his novel *Émile* with the statement that everything is good as it comes from the hands of the author of nature, and everything degenerates in the hands of man. He was talking specifically about children but the proposition was also meant to be general. From then on, nature began to be thought of as an intrinsic value, opposed to the values of society, thus forming a crucial part of the romantic ideal.

In the eighteenth century, until Rousseau, it was thought that nature existed to be improved upon. Wild, untamed nature was simply less agreeable than nature cultivated. The new spirit of Romanticism, on the contrary, both in Britain and all over Europe, demanded that the natural world of wilderness be regarded as awe-inspiring, a source from which men could draw understanding of their own being, and from which they would receive the most sublime aesthetic intuitions. The immutability and grandeur of nature, as well as its freedom from corruption, had the power to return men to their true identity, allowing them to see where they really belonged.

But alongside this new sensibility was already developing the thought that nature was the proper object of scientific study. Carolus Linnaeus, the Swedish botanist who died in 1778, had already established the binomial system of biological nomenclature that is the basis of modern classification. And, besides professionals, there were innumerable amateur diarists and watchers of nature, among them Gilbert White, whose *Natural History of Selborne* was read with delight by Charles Darwin as a boy. The idea was gradually taking root that, just as Newtonian physics provided laws governing the behaviour of all matter, so there were historical laws of development equally ineluctable in the biological world. The publication, in 1859, of Darwin's *Origin of Species* provided an explanation of how species evolved, by competition for life in a world of scarcity and the survival of the fittest. When Darwin published *The Descent of Man* twelve years later, the foundation was laid for a concept of human nature that, more or less, is with us still.

However, Darwin's theory of evolution was still less than explanatory because it was not clear to him how the variations within a species that led some to survive, others to fail, actually came about. For this, we had to wait for the birth of genetics in the twentieth century. So now we are Darwinians, but instead of concentrating on the macro scene – the development and behaviour of different species – our attention is fixed on the micro – the gene within, which determines the mechanism for this development.

Human nature is now tied up with nature as a whole, not simply by the bonds of romantic sensibility, nor by the human desire to observe and understand the natural world, but by the laws of biological science. Human beings are thought to be determined by their genes. But genes may be shared across the natural world, bringing all creatures, plants, fruit flies, men and women into a kind of universal cousinhood. So one may ask whether those who object to interfering with nature – especially those who object to genetic intervention in human subjects – are simply expressing their outrage at the diminution of the status of man, his now unspecial place in the universe. This would bring the objectors into line with the Church in the battles that raged in the twentieth century between T.H. Huxley, 'Darwin's Bulldog', and the theologians over the theory of evolution.

I do not think this is the explanation. That argument was primarily theological, and although Prince Charles expressed his view that biologists, in manipulating genes, were trespassing on areas 'which belonged to God and God alone', this did not mean that he was talking about theology. His argument could have been couched in terms of nature and nature alone without making any substantial difference. His words were metaphorical. The argument has moved on; for those who believe that interference with nature is wrong accept most of Darwin's premises. They are not, therefore, re-enacting the passionate disputes of earlier times. Why, then, the emotive word 'interference'?

I believe that those who use such terms feel both aspects of our idea of nature to be under threat. On the one hand, the romantic or aesthetic idea of nature is undermined by the threat of genetic manipulation. Nature in the Rousseauesque or Wordsworthian sense is untouched by human hand. It is wild in the sense that it is partly random, beyond total control. If crops are to be genetically engineered, animals caused to produce milk at all times, fruit made to last throughout the seasons, babies programmed to have the features their parents desire, then the element of astonishment (or disappointment) which is part of our reaction to nature will be eliminated. Our lives will themselves become unnatural.

Equally, our Darwinian ideas seem themselves under threat. In *Playing Safe*, Jonathon Porritt, former director of Friends of the Earth, wrote: 'The hard lines between different organisms and species are beginning to melt away. We can now pick and choose individual genes from one organism to introduce into a totally different and unrelated organism, crossing all biological boundaries in combinations that nature never could and never would bring together.' The laws of biological nature themselves, which we have accepted since Darwin, seem to be eroded. In a world where, we are constantly told, there are no universal laws of morality, how much the more terrible to be deprived of the certainties of natural law.

However, it is important to keep our heads, and not be swept away by the rhetoric of 'the natural'. We are indeed part of the natural world, though being brighter and more far-seeing than other animals, we have particular duties towards it, and towards our fellow humans. We should not lump things together in damaging dichotomies – the genetically modified or the organic, babies born incurably sick or designer babies. We must allow research to go on so that the techniques of genetic modification can be applied where good will come, while keeping our eyes open for unforeseen damaging consequences, including the enrichment of quacks or exploitative companies. If, for instance, it can be shown that the genetic modification of rice to make it more tolerant of adverse weather conditions would make a great difference to the level of nutrition in countries where rice is the basis of the diet, then common humanity demands that such rice should be made available. If it can be shown that stem cell transplant can effectively restore someone's damaged brain, then the common humanitarian concerns always at the heart of medicine should allow such treatment to be carried out. Genetic modification will not be all good or all bad, just as surgery or the breeding of specialist dogs and horses is not all good or all bad.

For my own part, I can think of only one development that I would think certainly wrong. Suppose that one day someone argued that if specific cells in the body could be renewed, then perhaps all cells could be, so that death itself could be eliminated or indefinitely postponed. I would regard this as unimaginably wrong. All art, all religion and all morality is built against the background of the fragility of life. Life is essentially ephemeral. But perhaps this is only an ultimate appeal to the rhetoric of nature.

Further reading

Alan Cromer: *The Heretical Nature of Science* (Oxford University Press, 1995)
Mary Shelley: *Frankenstein* (Signet, 2000)
Mary Warnock: *An Intelligent Person's Guide to Ethics* (Duckworth, 2000)
Lewis Wolpert: *The Unnatural Nature of Science* (Harvard University Press, 2000).

Can We End Disease

? Until the 1980s, it looked as if the West at least had disease on the run. The twentieth century was full of new treatments and vaccinations as test-tube babies followed heart transplants followed penicillin. And then, wham – it was tombstones and doctors in protective suits as panic set in at the thought of a new deadly, incurable disease – Aids. Since the eighties, we have lurched from medical nightmare in the form of the new variant CJD to genetic breakthrough. Although it is still early days, the trumpeting of the decoding of the human genome and reports of work on the human proteome promise a cure within decades for all sorts of diseases, including Alzheimer's and some forms of cancer. Already the sequencing of various diseases has led to the development of new drugs. Then there are the possibilities being held out by stem cell therapy – which could include the eradication or dramatic reduction of Parkinson's, diabetes and heart disease, among others – gene therapy and genetic cloning.

It is hard to believe, with all this talk of progress, that the modern idea of disease and infection is not even two centuries old. Although the microscope was invented and microbes first identified in the seventeenth century, it was only in the nineteenth that scientists realised that microbes caused disease. Before that, a variety of explanations existed, ranging from the belief that disease was a punishment from God, or the gods, to the idea that it was related to foul-smelling air or miasma.

A number of scientists paved the way for the breakthrough nineteenth-century discoveries of Joseph Lister, Robert Koch and Louis Pasteur. They include Edward Jenner, pioneer of immunisation through his work on smallpox, and Rudolf Virchow, founder of the modern concept of the pathological processes of disease, under whom Koch studied in Berlin in 1866. But it was Koch's work on the transmission of disease by the blood, his technical breakthroughs for cultivating and tracking bacteria and his discovery of the bacillae for tuberculosis and cholera, Pasteur's germ theory of disease and his development of vaccines for rabies and anthrax, and Lister's work on preventing infection during and after surgery, which together provided the foundations of modern medicine.

In the twentieth century, the fast pace of discovery quickened. With the basics of infection revealed, scientists could concentrate on the next stage – preventing and treating it. Since the early twentieth-century discovery of penicillin, we have developed a range of antibiotics to treat disease, as well as a host of vaccinations to ensure we never get sick in the first place. We have witnessed the eradication of diseases that killed huge numbers of people, such as smallpox and polio. And we have seen sweeping technical advances in surgery for the treatment of such conditions as heart disease and cancer. The current focus on the genetic causes of disease, begun with the work of James Watson and Francis Crick on DNA, promises a future of sophisticated, individually tailored medicine.

But this emphasis on the individual does not deny the social and environmental influences on health. Pollution, the explosion in tobacco use, radiation, pesticides and food additives have all gone under the microscope, and research is concentrating on their interaction with a number of factors, including genetic ones. In many cases, research results are still unproven, or not straightforward. For example, asthma has been linked with pollution, but recent research has concentrated on evidence showing it may be caused by an environment that is too clean. And then there are fears of the possible health effects of genetically modified food.

There has also been a resurgence of interest in the impact of social deprivation on health, with poor diet – and the long-term effects of poor maternal diet on the foetus – being a major area of research. In the United Kingdom, the widening gap between the very rich and the very poor in the last two decades, demographic changes which put greater strain on the health service due to an increase in the number of elderly people needing treatment, and lack of investment in the health service have created pockets of the country where diseases connected with poverty are significantly higher than in others. These are also usually the areas which tend to lose out in so-called 'postcode lotteries' of access to the best treatment.

Worldwide, health inequalities have meant that there has been little change since the nineteenth century in the type of diseases affecting developing countries, and relatively little research into the way different diseases affect poorer population groups. Despite progress in the West, essentially the same top ten infectious diseases – bar Aids – account for the major burden of disease in the developing world. For example, malaria is estimated to kill at least one million a year, with 90 per cent of deaths in Africa. Tuberculosis, in many cases linked to Aids, kills around 1.9 million a year, almost all in developing countries, where resistance to treatment is growing. And the future outlook is not positive.

According to Médecins Sans Frontières (MSF), an international emergency medical relief organisation, the increasing privatisation of drug development and the public sector's failure to take a needs-based approach to it have led to neglect of the diseases of the poor. Moreover, the cost of development has been pushed up by increasing demands from Western consumers for higher standards of testing and concerns about drugs' side effects. At least drugs exist for Aids, even if current patent laws and market forces mean they are too expensive. For many other diseases that affect developing countries, few new drugs have been developed in recent years. For example, sleeping sickness, which affects around 500,000 people, was until recently treated with an arsenic-based medicine. MSF argues that there needs to be a global research and development agenda, reliable long-term funding and a global treaty obliging pharmaceutical companies to reinvest a certain percentage of their profits into researching neglected diseases.

The argument is not merely ethical but practical since increasing globalisation means that there is no longer an easy divide between rich and poor countries. The UK, for example, has seen a sharp increase in the number of TB cases in recent years, most having their origins in developing countries. Research on immigrant communities, often underfunded, suggests health inequality may not be the only factor at play in the relationship between globalisation and disease. For example, Robert Wilkinson, a Wellcome Trust fellow, has found a link between the facts that Gujeratis living in London have lower vitamin D levels than those living in Asia and also have higher rates of reactivation of TB.

Would eradicating poverty rid the world of disease? It would certainly lessen its impact, but wealth and 'progress' create their own health problems. The fact that we live longer has led to a growth in diseases associated with old age. An over-rich diet, full of saturated animal fat, has brought more heart disease and cancers. And things look unlikely to improve – researchers say one in three eleven-year-old girls in the UK is overweight and one in ten is clinically obese. Allied to this is the growth in sedentary jobs and a general decline in exercise, coupled with the stresses associated with the instability and fast pace of modern life, including increased family breakdown and moves to a twenty-four/seven way of life. While stress is still unproven as a cause of disease, it is acknowledged as a factor in depression, which the World Health Organisation predicts will become the second greatest cause of the global disease burden within the next twenty years. There is also agreement that stress can be a factor in undermining the immune system. Indeed, concerns about boosting the immune system, whether through diet, alternative therapy or other methods, have been an increasing focus for scientists in recent decades, with Aids bringing an extra urgency to the field.

Moreover, like globalisation, other conditions associated with development have speeded up the chain of infection. Urbanisation, for example, has provided effective breeding grounds for infection chains, and global warming has led to fears of serious outbreaks of infectious disease in the north due to changing migration patterns for disease carriers such as mosquitoes.

Sometimes, too, science can create its own monsters. Over-prescription of antibiotics has led to fears of the development of antibiotic-resistant bacteria as some microbes manage to evolve to withstand their effects. These new 'superbugs' become more prevalent because the antibiotics kill off weaker competitors, in turn spawning new diseases such as MRSA, methicillin-resistant *Staphylococcus aureus*. Existing diseases are also proving increasingly difficult to treat. The virus that causes Aids and the bacillus behind TB, for example, are both evolving at a rate which chemists are finding difficult to match. The failure of patients to stick with a course of treatment has also helped the rise of drug-resistant strains.

Far from eradicating disease, affluent societies have also seen the birth of new mystery illnesses in recent years, such as chronic fatigue syndrome. Is it a psychosomatic illness, a new disease or, as some sufferers believe, a strain of an older one – poliomyelitis? Symptoms include extreme physical weakness and painful muscle spasms, but, although it is now officially recognised as an illness by the government, some doctors are still sceptical. Some parents of sufferers have even been accused of child abuse because they will not accept that their child needs psychiatric treatment. Another new mystery disease is Gulf War syndrome, whose existence is still disputed in medical circles. Several theories about its causes are being investigated, including the cocktail of drugs given to soldiers serving in the Gulf War and the role of depleted uranium and organophosphates, but the jury is still out. And then there are the diseases which time forgot – anthrax, bubonic plague, smallpox. These are diseases we thought we had defeated but kept for military purposes. They may now come back to haunt us in the form of biological warfare – perhaps in an even more destructive form since scientists have been manipulating them to form superstrains. There is also talk of genetically engineering new germs for use in war.

On the one hand, then, we have come from ignorance about the causes of disease to an ability to manipulate disease for our own ends, even if we cannot ultimately control it. But at the same time, an increasing distrust of science and medicine has grown up, stemming from a range of issues including concerns over scientists 'playing God', ignorance on the part of ordinary people of how science functions, changing attitudes to authority and the impact of BSE or 'mad cow disease'. This has resulted in a huge new interest in alternative medicine, with its concentration on balancing the body and the mind. It harks back to earlier pre-nineteenth-century ideas of disease treatment, as well as responding to panic reactions to medical or other scares, such as concerns over the mumps, measles and rubella (MMR) vaccination.

Nevertheless, there has undoubtedly been huge progress in our understanding of disease and our ability to treat it, although we are still a long way from understanding how all the many factors involved in disease work. Many scientists no longer believe that science can provide all the answers. There is an urgent need for academics from a variety of backgrounds to come together and look for solutions and for governments to recognise the need for global public investment in tackling both the major cause of disease and the major obstacle to its eradication – poverty.

Mandy Garner
Features editor of the Times Higher Education Supplement

Big Questions in Science

Can we end disease?

John Sulston

Former director of the Sanger Centre, Hinxton, Cambridgeshire,
where he led the UK contribution to the Human Genome Project

By the standards of some poor communities, the prosperous sector of the world has abolished most disease already. Thanks to developments in public health and antibiotics, in cancer treatment and heart surgery, we have gone from disease being normal to disease being abnormal. But people continue to be ill, and it is of no comfort to them to say it is abnormal. Therefore, much effort continues to go into tackling the problem.

Sequencing the human genome has been a striking and hyperbolically reported step forward in this effort. Just as important from the point of view of disease is the sequencing of genomes of pathogens (bacteria, viruses and other parasites) and of other animal genomes that will help us to understand our own. From a long-term perspective, the excitement is justified by the fundamental importance of collecting these codes. But what can we actually expect from them in the foreseeable future?

Pathogen sequencing will open up new targets for drug and antibody attack. This is important not only for solving old problems but also for dealing with the increasing threat of antibiotic-resistant strains that are appearing all the time.

In human genomics, the emphasis is now on analysing people's variants of the code and comparing them with individual health and medical problems. Widespread genetic testing brings with it issues of privacy and human rights that must be dealt with, but the benefits are great. Interaction of genetic variants with diet and other aspects of lifestyle will become well known, enabling people to be given accurate advice. Diagnosis of hereditary disease will become increasingly precise and common-place as quick and cheap testing methods are developed. So will diagnosis of susceptibility to drug treatments and their side effects. This means that, at least in critical cases, treatment may be made more efficacious, either by helping the selection of existing drugs or by the design of new ones. Some feel that, in the early years, these areas of pharmaco-genomics and pharmaco-genetics are likely to have a greater influence on clinical practice than anything else.

Accurate diagnosis opens up the possibility of pre-natal selection through selective implantation or abortion. This will allow serious hereditary disease to be avoided in the future, which in itself is immensely desirable. But we must decide what is allowed, what is desirable and what is abnormal. Society is now grappling with the ethics of embryo choice. On one side, we have people who argue that these activities are wrong, under any circumstances. On the other, we already have children who are suing their parents for allowing them to be born disabled. As more genetic markers become known, the boundaries of permis-sible selection will have to be drawn ever more clearly in order to protect medical practitioners. Some parents will want selection to be very stringent and even extended to quantitative traits, such as intelligence. Conversely, many disabled people campaign for the absolute rights of the unborn, on the understandable ground that under more extreme criteria of normality they would themselves have been aborted.

Great efforts are going into gene therapy – the replacement of a faulty gene with a good copy. This is diffi-cult because of the problems involved in delivering the healthy gene to the cells that need it and of having it work correctly and stably. But it is beginning to work for diseases of the immune system, where cells are relatively accessible to manipulation, and there will be many more successes in the future.

One of the most terrifying diseases for us nowadays is cancer. While many other causes of death have been eliminated, this still strikes while we are in our prime. Great strides have already been made in treatment, but it is reasonable to suppose that detailed genetic analysis of tumours, leading to accurate targeting of toxic drugs on to the offending cells, will lead to a surge of improvement in the next decade or so.

The most important aspect of reading the human genome is that it is a key step towards total understanding of our bodies. Possession of the code is not so much a problem-solver in itself, but a resource, a reference to aid and guide research on the systems of the body. This is why it must be freely available to everyone. Physiological pathways are vastly complex, and their understanding requires insight and experimentation far beyond the code itself. Access to the basic instructions both empowers and sets limits on the hunt.

Incidentally, the very complexity of these systems, and of their controls, refutes doubts expressed about how far genes are significant. Such doubts surfaced following the revelation that our genes number only about 30,000–40,000. But genes have multiple interactions, and it is the combination of their variants – an inconceivably large number – that has to be considered in evaluating the possibilities of the human body, not simply the number of genes taken one at a time.

Greater understanding of our bodies prompts new questions. Old age itself can be a disease, for many the most devastating of their lives. Progress in prolonging life is no good without considering its quality, so there is tremendous interest in tackling senility in all its aspects. But what of death? Is that a disease too? And do we want to abolish it? For me, the answer is absolutely not. I find that I am increasingly impressed by people younger than me rather than by those my age or older, who tend to be regarded with exaggerated respect if they have achieved a measure of notoriety. There should be the strong sense that it is good to move aside and let a new generation take over. Yet for some immortality is an irresistible dream. How might it be achieved?

One line of thought is reflected in the current

fascination with the idea (or indeed the practice) of human cloning. But genetic cloning, quite apart from practical and ethical difficulties, is in no way a recreation of oneself as an individual. To clone oneself is to give rise to one's identical twin – but a twin born in another age, with its own pressures and ideas, and so a new personality. More logical, though as yet fanciful, is the notion of scanning one's thought processes into a computer that controls a robot. The transferred mind would be, in principle, much more nearly akin to a replica of the personality than would a genetic clone.

Should we keep the bodies we have, how are we going to fix them if diseased? Methods of repair will become ever more subtle and perfect. Gene manipulation will not be the only product of the Human Genome Project. The consequences of really understanding how our bodies work will mean that soon there will be no gap in the spectrum from surgery to biochemistry. Every ailment will be tackled by a combination of tools, and prosthetic devices will be ever more commonplace, effective and unobtrusive.

But how much non-biological hardware can we hook up to a human body and still call it human? This is no joke. Once the prosthetic researchers have solved the problem of making a reasonably stable connection between the nervous system and computers – and we are approaching that point – then surely there will be a demand for various sorts of brain extension. We are good at learning to use tools, and the experiences of those who have cochlear implants or play with virtual reality devices is an early indicator of what is to come. A little more memory perhaps? More processing power? Why not? And if so, perhaps a kind of immortality is just around the corner.

All this discussion is largely from the perspective of the rich. The endless struggle with pathogens is much more apparent in the developing world, where they are still the major cause of disease. The proportion of research funding spent on tropical diseases is minute. The most profitable products for the pharmaceutical companies are for the treatment of depression, high cholesterol and gastric ulcers. There is no profit in drugs for poor people. Unless this situation can be corrected, the world will be not only unjust but also immensely unstable. Market forces alone will not be able to level the global playing field. It demands a deliberate search for justice. One contribution to the search is the free release, by the Wellcome Trust and its partners, of genome sequences, which ensures that this fundamental information is available to all.

There are signs of growing inequality within rich countries too. Everyone knows, although it is rarely discussed, that there will be increasing difficulties with the equitable distribution of healthcare. Again, conscious democratic decisions will be required if the benefits of new technology are to be shared rather than auctioned.

To end disease is an admirable objective. I think a more urgent one, though, is to distribute more fairly the expertise that we already have. Life will continue to improve as we gain more and more understanding and keep trying to fix things. The dangers ahead are from inequality much more than from a failure to solve medical problems.

Further reading

E.H. Ackernecht: *A Short History of Medicine*
    (Johns Hopkins University Press, 1982)
Roy Porter: *The Cambridge Illustrated History of*
    Medicine (Cambridge University Press, 2001).

Will We Ever Be Pain-free ?

'For all the happiness mankind can gain/Is not in pleasure, but in rest from pain,' wrote the poet John Dryden. So he would surely be overjoyed that the past three hundred years have delivered the more privileged among us to a freedom from pain unimaginable in the seventeenth century.

Medical science may not quite have abolished the experience of pain, but it seldom questions the rightness of this as a goal. Contrast that with the attitude of our forebears, many of whom might have been less inclined to ask *how* we could abolish pain than whether we *should*.

Throughout most of human history, pain has been regarded as divine retribution or as part of the natural order of things. Enduring it has been held to confer virtue. In a time when pain control was limited or non-existent, such attitudes were understandable: they made the best of an experience that was not only unpleasant but also inescapable. With the advent of effective pain control, these views might have been expected to disappear. Certainly they have faded, but they have not entirely gone. The late Patrick Wall, Britain's most celebrated pain researcher, quotes Pope John Paul II: 'What we express with the word suffering seems to be particularly essential to the nature of Man. Sharing in the sufferings of Christ is, at the same time, suffering for the Kingdom of God . . . Suffering contains, as it were, an appeal to Man's moral greatness and spiritual maturity.'

Wall points out that the Pope seems not only to accept suffering but to glorify it. He adds: 'This powerful statement has had practical consequences in Catholic countries, particularly in treating terminal cancer, where some doctors have hesitated in their treatment of pain and suffering because the treatment might intrude on the patients' acts of redemption.'

Not that you have to go to Catholic countries, or indeed rely on the prompting of priests, to find physicians who have pondered the probity of pain relief. In the 1830s James Simpson's introduction of chloroform as an anaesthetic during labour prompted suspicion and even hostility. Fears about its safety may have been justifiable, but many argued that pain during labour was somehow part of the natural order of things, and not to be ameliorated. Churchmen waded into the argument by quoting Genesis: 'In sorrow thou shalt bring forth children.' The debate was not settled for another twenty years, and then it was by royal rather than medical prerogative. Queen

Victoria accepted chloroform during the birth of Prince Leopold – and that was that.

The notion that limiting pain is unnatural comes unstuck when you appreciate that nature itself has evolved pain-limiting mechanisms. Time and again men injured in battle report being unaware that they had been seriously wounded, or indeed injured at all, until the conflict had abated. Only then did the pain begin.

The survival value of such arrangements is clear enough. Pain is a message designed to alert the organism that something is wrong. But for a caveman running from a sabre-toothed tiger that has bitten off part of his arm, a pain signal would be irrelevant. Indeed, it would be counter-productive; excessive awareness of the injury would simply hamper his efforts to stand and fight, or turn and flee.

The pain-modulating system is sophisticated and not yet fully understood, but the body is known to manufacture molecules called endorphins, which are natural counterparts of the analgesic (anti-pain) drug morphine. Their discovery offers one explanation for soldiers' toleration of battlefield wounds: they are producing their own painkillers. It also explains why drugs such as morphine are such effective analgesics: they work because they are making use of the body's own pain-suppression mechanism.

That is not all. In trying to account for the variability of an individual's pain responses from one occasion to another, Patrick Wall and his McGill University collaborator Ronald Melzack devised what they called the 'gate theory of pain'. While some nerves convey pain signals to the brain, others carry messages back from it. These latter can interfere with incoming signals and reduce or block them – close the gate, to use Melzack and Wall's terminology. Conversely, by opening the gate more widely, awareness of pain can be heightened. It was exploitation of this idea that led to the method of pain control known as transcutaneous nerve stimulation. A small electrical stimulus, applied to the correct area of the skin, can be used to close the gate artificially.

Although relief of pain is one of the routine functions of the modern doctor, it played relatively little part in medicine prior to the Renaissance. Not that there was much physicians could have offered anyway apart from mandrake, henbane and alcohol. The advent of opium represented a major step towards more effective analgesia. But it was the nineteenth century that witnessed the first development of our most familiar drugs, including morphine, codeine and aspirin. With the invention of nitrous oxide, chloroform and ether, surgeons no longer needed to equate skill with speed, and patients were spared the ordeal of remaining conscious during amputations and other even more risky operations.

One of the hurdles to good pain control is its subjectivity. Doctors can measure blood pressure, cholesterol or sugar in urine and medicate accordingly. But when it comes to pain they can only ask the patient. My 'agonising' may be your 'slight', and the tendency of medical staff has been to underestimate severity. One remedy, particularly for dealing with post-operative pain, has been found in patient-controlled analgesia (PCA).

A patient using this technique is fitted with a motorised syringe that delivers a small dose of pain-killing drug into a vein at the press of a button. Those who want more pain relief simply press the button more often. The system includes various safeguards to prevent accidental overdosing, and is widely rated a great success. Experience shows that, among a large group of patients using PCA, the overall consumption of drug remains about the same as it would have been using conventional prescribing. What does alter is the distribution. Some patients take more, some less.

Treatment of pain has also been plagued by the assumption that it must have a physical origin. Professionals and lay people have both clung to this mistaken view. And patients who are told that the pain is all in the mind are apt to interpret this as a polite way of being told they are malingering.

Most pain does have a physical origin, whether in the form of a burn, a microbial infection or some other injury, but not always. Chronic afflictions with no apparent cause may begin with an acute response to physical injury that is inappropriately maintained. The injury itself and all observable signs of damage heal but the pain state set up within the central nervous system lingers on. Psychiatrist Andrew Hodgkiss of St Thomas's Hospital in London argues that the medical profession must bear part of the blame for the view that pain must have a physical cause.

For him, 'Pain is a complex lived experience of human beings with roots in their autobiography, their emotional state and their concentration, as well as their sensory input. If we hadn't discarded all those ideas, and gone instead for a simple relationship between disease and pain, we wouldn't be having to struggle with saying it's in the mind. Of course pain is in the mind. It's a lived experience, it's a perception, it's an emotion, it's all these things at once. All pain is in the mind even if there's a great big lesion.'

The futility of continuing to search exclusively for a physical cause when none can be found has important implications for treatment. It has led to the creation of a handful of last-resort pain clinics that place the emphasis not on eliminating pain – though they continue trying to do so – but on helping patients to live with it. The approach relies broadly on cognitive behavioural psychology.

The first aim of these clinics is to restore patients – many of whom will have become inactive on account of their pain – to physical strength and health. Then comes a programme of practical advice on how best to live as normal a life as possible without exacerbating the pain. And then there is the more formal psychological help. It rarely removes the pain entirely but it does help most people to cope more effectively.

Do we really want to live in a world free of all pain – any more than we seek to live in one free of risk? So addicted are we to the latter that we flock to theme parks to simulate the experience of danger, we circumvent the efforts of officialdom to keep us safe, and we even invent and pursue unnecessary but genuinely risky pastimes, from rock-climbing to parachuting. So do we really want to abolish all pain? Certainly the masochist does not. And even those without the stomach to endure pain for themselves may turn up to see other people doing so. Performance art, which reached its apogee in the seventies, has featured artists who variously shoot, cut, trepan, pierce, scarify and burn themselves – and hold an audience while doing so. Clearly we are intent on eliminating pain from medicine. But eliminating it totally? Maybe not.

Geoff Watts
Writer and broadcaster on science and medicine

Will we ever be pain-free?

Ronald Melzack
Emeritus professor of pain studies at McGill University

'Stop pain' has become a rallying cry for people who seek improved care for countless patients suffering agony due to cancer, arthritis, nerve injuries or other causes. Severe, prolonged pain destroys the quality of life of those who suffer it and the need to abolish this kind of pain is urgent.

However, we must first recognise another kind of pain – one that has positive aspects. Certain brief, acute pains, usually suffered after an injury or infection, have genuine survival value. These pains make us rapidly pull a hand away from a hot oven, lift a foot off a sharp object, or telephone for an ambulance when we develop sudden discomfort in the abdomen or chest. Such rapid responses to pain are aimed at preventing or minimising serious damage to the body and are important for learning to avoid future encounters with objects and situations that are dangerous. Heightened sensitivity to pain during healing also prevents us from reinjuring ourselves and postponing full recovery.

People who are born without the ability to feel pain provide convincing testimony of how valuable some acute pains can be. Many of these people sustain extensive burns, bruises and lacerations during childhood, and have difficulty learning to avoid inflicting serious wounds on themselves. Failure to feel pain after a ruptured appendix, which is usually accompanied by severe abdominal pain, led to near-death in one such man; another walked on a leg with a cracked bone until it broke completely. Similarly, a woman sustained numerous cuts and burns without feeling them, and her mouth was scarred from blisters as a result of drinking burning-hot liquids. Her daughter had the same condition and, aged seven, pressed her buttocks up against a grated bathroom heater after taking a bath, branding herself with five large cross-hatches without feeling any pain. Such stories make it clear that we do not want to be totally free of the ability to feel acute pain. It saves lives.

In contrast, chronic pain is destructive and has no redeeming features. One form of chronic pain is associated with unremitting diseases, such as cancer and arthritis, that destroy body tissue and with various well-defined pathologies of the body's functions, such as protruding discs in the spine or insufficient blood supply to heart tissue. Attempts to relieve these pains often work. Cancer pain, for example, can be greatly diminished – sometimes even abolished entirely – by appropriate doses of morphine or other opioid drugs, often given in combination with other classes of drugs, or even with psychological therapies, to enhance and maintain their effectiveness. Yet, despite the best efforts of major hospitals with outstanding facilities, between 5 and 10 per cent of patients with cancer continue to have moderate to high levels of pain. Near the end of life, when they serve no useful purpose, these unrelenting pains make living unbearable and have been known to drive some patients to suicide. Will we ever be free of them?

It is possible. New drugs are in the pipeline and, with luck, cancer pain may be whittled down by the development of specialised, selective drugs or other therapies. The explosive growth of research in the field of pain in recent years in both the laboratory and clinic has led to the discovery of two new classes of very effective analgesic drugs. These were never suspected of having this kind of power but were developed to control epilepsy and psychological depression. The anti-epilepsy drugs are usually used to control neuropathic pain, which is associated with pathology of peripheral nerves, while the antidepressants relieve several kinds of pain even in patients who are not depressed, so that their analgesic action is separate from their effects on depression. Powerful new drugs to relieve arthritic pain have recently become available, and even more powerful ones seem to be on the way. It is possible that chronic pains associated with major, definable pathology in the body may eventually yield and become pain-free zones in the field of medicine.

Unfortunately, this optimistic outlook does not apply to the kind of chronic pain that has no well-defined cause, such as phantom limb pain. People who undergo amputation of an arm or a leg almost immediately feel a 'phantom limb', which seems so real that they sometimes try to step off their bed on to their phantom foot or reach for a ringing telephone with their phantom hand. The reality of the phantom is especially vivid in the 60 to 70 per cent of amputees who suffer terrible pains – cramping, burning, crushing – in calf, ankle, hand or other areas. The pain may be relieved briefly by local anaesthetic injections or some analgesic drugs taken orally. But, tragically, the cause of the pain is poorly understood and no effective treatment has yet been found.

Neurosurgeons may remove the little ball of tangled neurons (the neuroma) that form in the stump and pain may be relieved for a while, but it soon returns. Then a neurosurgeon, often urged on by the desperate patient, may cut the nerve near its point of entry into the spinal cord, and later, may cut part of the spinal cord, usually without prolonged relief. Sometimes electrodes are lowered into the brain to burn out parts of the thalamus or nearby areas. Even the cortex representing the missing limb may be removed. None of these operations has been effective long enough to be counted on as a reliable therapy. Even when a few inches of the spinal cord are completely removed, leaving a gap so that no information from the lower limbs can reach the brain, the terrible pain (it had to be terrible for this major surgery to be done) usually persists in the same areas of the phantom body.

The reason is now at least partly clear. When sensory input to brain cells is cut off, these cells begin to fire spontaneously in abnormal bursts. The pathological patterns of nerve impulses are assumed to generate pain perception. Anti-epileptic drugs help some of these patients, but most continue to suffer. Again, new drugs are being developed, and some of them, in the future, may work. Sadly, there are no grounds for optimism that a drug to control this horrendous, unrelenting pain is imminent.

But at least these types of pain are beginning to be better understood. A striking feature of chronic pain, which includes most kinds of backache, headache, facial pains, musculoskeletal pains, pelvic and visceral pains, is that it involves activity in all of the brain. Not only are the sensory receiving and processing areas of the brain known to be active when people are in pain, but areas of the limbic system (which plays a role in emotion, motivation, suffering), and cognitive areas (which evaluate the situation, give meaning to it, foresee hope or doom) all play a role too.

Take headaches. Two of the most common forms are tension headaches and migraine (vascular) headaches. Close study shows that tension headaches are not always associated with increased tension in head and neck muscles, and migraine headaches may come and go with no clear relationship to blood-pressure changes in the vascular system of the head. Also, stress plays a major role in both kinds of headache but not in a predictable way. All three factors – muscle tension, blood flow and stress – are important to varying degrees but no one of them alone is a reliable predictor of the type of headache or when it will come and go. That is why treating only one of these aspects rarely seems to be enough to make a substantial difference for a long enough period of time. Effective treatment often involves all three.

While we have excellent new drugs for some kinds of headache and our understanding of them has advanced considerably in recent years, we cannot cure them all and continued suffering by millions of people indicates we still have a long way to go. Yet there is still room for optimism. Now that we recognise the powerful role of the brain and all its contributions – sensory, emotional, cognitive – we have recently begun to explore the brain's mechanisms for generating consciousness and the perception of pain and suffering. Indeed, we may be at a stage in our science comparable to Copernicus's proposal in the fifteenth century that, contrary to common sense, the Earth rotated around the Sun. Later confirmation of this simple fact led, in less than five hundred years, to the discovery of our solar system, then eventually to our vast, expanding universe. So too, contrary to common sense, we now know that the brain can generate pain in the absence of any evidence of injury, infection or other pathology. The brain is extraordinarily complex with a hundred billion nerve cells and trillions of connections but scientific progress is certain eventually to reveal its secrets. There is certainly the hope that those secrets will also illuminate ways to wipe out the horrible pain and suffering that afflicts people throughout the world.

Further reading

Ronald Melzack (with Patrick D. Wall): *The Challenge of Pain* (Penguin, 1999), 'The Tragedy of Needless Pain' (*Scientific American*, vol. 262, 1990, pp. 27–33), 'Pain and Stress: a New Perspective' in R.J. Gatchel and D.C. Turk (eds), *Psychosocial Factors in Pain* (Guildford Press, 1999, pp. 89–106)

Patrick D. Wall and Steven Rose (eds): *The Science of Suffering* (Weidenfeld & Nicolson, 1999).

Can We End
Hunger ?

? At the 1996 United Nations World Food Summit, 186 countries pledged to halve the number of undernourished people by 2015. According to the UN Food and Agriculture Organization, the target is unlikely to be met. Despite all our world's technical advances, we appear unable to feed ourselves. In sub-Saharan Africa, for example, according to the International Food Policy Research Institute (IFPRI), one-third of children continue to go to bed hungry and 'have their mental and physical development compromised by the ravages of hunger'.

Of course, people have gone hungry throughout history. Perhaps the earliest mention is the biblical story of Joseph in which famine grips the land now called Palestine. People have also been trying to understand patterns of hunger for centuries. In 1798, clergyman and scholar Thomas Malthus suggested that while food production grew arithmetically, population grew at a geometric rate and hence outstripped the means of production, which inevitably limited population growth. Esther Boserup, writing in 1965, was more upbeat. She suggested that demographic pressure promoted innovation and higher productivity, hence, in contrast to Malthus, argued that food production could keep pace with population growth.

In his 1981 book *Poverty and Famines: An Essay on Entitlement and Deprivation*, Nobel Prize-winning economist Amartya Sen showed that famines had occurred in recent history even when the supply of food was not significantly lower than during previous famine-free years. He challenged the view that food shortages were the most important explanations of famine. Twenty years on, most scholars agree that there is currently sufficient food globally to feed the entire world population of six billion, with many countries producing food surpluses. Hunger has become far from a simple production-related issue.

The green revolution of the 1960s, 70s and 80s, which involved the introduction of high-yielding varieties of plant, in addition to fertilisers and pesticides, saw significant increases in production in some developing regions. Artificial fertilisers had been introduced in the 1950s, but scientists found that they made traditional wheats grow so tall they fell over. American agricultural advisers working in Japan after the Second World War identified dwarf wheat varieties that were brought back to the United States and formed the basis of the plant breeding associated with the green revolution. These plants, as well as not falling over, put less photosynthetic effort into producing stems and more into grain production. A similar cross-breeding programme was undertaken with rice. As a result, the number of malnourished people has dropped from one billion in the 1970s to 800 million today, a decline from 37 to 18 per cent.

But increased productivity has not been felt in every developing geographical region, nor uniformly for everyone within each country. Sub-Saharan Africa, for example, has tended – perhaps because of the distinctiveness of crops and farming methods – to miss out on green revolution technical developments. As a result, per capita output has declined since the 1960s, with associated hunger. Tim Dyson, professor of population studies at the London School of Economics, suggests that widespread political instability, ethnically heterogeneous nation states, neglect of agriculture by governments and extremely rapid population growth are also to blame. Even within individual nations, there can be both food surpluses and hunger. India, for example, has tens of millions of tons of stored cereals, yet is home to one-third of the world's hungry.

These people, explains Jules Pretty, professor of environment and society at the University of Essex, are hungry 'because they are poor'. They do not have access to food nor to the resources to buy food. What is needed, he says, are the right sustainable, pro-poor policies. With between 70 and 75 per cent of the world's hungry in rural areas, says Per Pinstrup-Andersen, director-general of the IFPRI, the best approach is to increase productivity on small-scale farms by giving farmers access to whatever tools they need. Others stress the need for strategies to increase economic growth and improve health and education for the poor in both rural and urban areas.

While current strategies to end hunger focus on such political and geographical solutions, science still has a role. World population is predicted to rise by a third to eight billion by 2030, and will coincide with increased urbanisation, an ageing population in some areas, and changing tastes: people will consume more meat – a less efficient use of cereals – as economic prosperity grows. In addition water scarcity, soil degradation and climate change, possibly causing desertification, could affect farmers' production capabilities.

Anthony Trewavas, professor in the Institute of Cell and Molecular Biology at the University of Edinburgh, and Oxford University's

Christopher Leaver argue that over the next fifty years food production will have to double or triple on the same area of land, with decreasing water supplies and without damaging the environment. 'If we keep technology at 1993 levels, to feed people in 2025 we will have to plough up another 800 million hectares of forest,' says Trewavas. 'But we don't have huge new lands.' He argues that further improvements in agricultural technology are needed to increase crop yields.

Over the past twenty years, new understanding of how cells and organisms work has revolutionised plant science. Where conventional plant breeding relies on crossing similar species to bring about hybrids with desired traits and is constrained by the traits already contained within a species and its close relatives, scientists can now select one or more genes from a species and transfer them directly into another species. They can even take genes from other organisms such as bacteria and introduce them into plants.

This technology allows scientists to make specific plant processes more efficient by introducing genes to improve yields, for example. Or they can introduce genes to increase resistance to pests and pathogens, which currently kill 40 per cent of crops, or to abiotic stresses such as heat and drought.

Worldwide, 44 million hectares of transgenic crops were grown in 2000. This first generation of transgenics, under cultivation mostly in the US, but also in significant quantities in Canada, Argentina and China, has involved relatively straightforward manipulation of single genes to confer resistance to herbicides and insect pests. But it has caused huge controversy in Europe where consumers are seen to benefit little and where there are concerns for the environment and human health. The worldwide environmental pressure group Greenpeace, for example, opposes all releases to the environment of genetically modified organisms.

The balance of benefits and risks may be very different for transgenic crops in the developing world, says Gordon Conway, president of the Rockefeller Foundation, a US charity which over fifteen years has put $100 million into plant biotech research. He calls for a doubly green revolution, 'that repeats the successes of the old but in a manner that is environmentally friendly and much more equitable' (the Rockefeller Foundation estimates that of the $2.5 billion spent on research and development biotechnology for agriculture, just $75 million is spent in the developing world). This new revolution, he says, should include applying modern biotechnology 'to help raise yield ceilings, to produce crops resistant to drought, salinity, pests and diseases, and to produce new crop products of greater nutritional value'.

Perhaps the most interesting GM achievement so far has been the introduction of genes that produce beta-carotene – the precursor to vitamin A – into rice grains. Beta-carotene is present in the leaves of rice plants but conventional plant breeding has been unable to put it into grain. Scientists in Zurich have done this using one bacterial and two daffodil genes, producing a rice plant – Golden Rice – with the potential to provide sufficient beta-carotene to meet human vitamin A requirements. Currently 150 million children a year in the developing world lack vitamin A, leading in some cases to permanent blindness and even death.

Inserting specific genes through genetic engineering can achieve in a few years what could take decades with traditional plant breeding. Florence Wambugu, one of Africa's leading plant geneticists, who is working on a biotech solution to sweet potato virus, has suggested that GM may also work better for Africa than green revolution technologies because all the technology to control insects is packed within the seed. There is no need to educate farmers on the use of fertilisers, for example. She is concerned that antipathy to GM in Europe, where transgenic crops are not needed to stop hunger, could stall its introduction in Africa.

Again, therefore, it becomes a political issue. For Pinstrup-Andersen, what is lacking is not understanding of how we can end hunger but the political will to do it. 'Those in power don't have malnourished children,' he says. 'They are not giving food a high enough priority.'

Julia Hinde
Freelance writer on science and education

Can we end hunger?

Brian Heap

*Vice-president and foreign secretary of the Royal Society, master
of St Edmund's College, Cambridge and special professor,
University of Nottingham*

The last half-century showed real progress as food
production outpaced population growth in many
regions of the world. Cereal production increased
arithmetically from well under 300 kg per person in the
1950s to over 350 kg in the 1980s. Since then, world grain
output has fallen behind demographic growth, not because
of any sudden spurt in population growth or a genetic
ceiling in plant breeding but primarily because of the effect
of economic curbs on production by major producers in
North America to maintain cereal prices. Demographic
growth *has* overwhelmed cereal production in sub-Saharan
Africa, however – the only region since the 1970s to be
tortured by major famines.

Food production can be increased and malnutrition
reduced through many avenues – appropriate technologies
derived from existing and new knowledge, improvements
in education and women's status relative to men, local
innovations by farmers themselves, and national and international policies that promote trade, regulate food prices
and exchange rates, and improve access to markets. To
date, global trade in farm products has been liberalised
only slowly, with tariffs on manufacturing goods falling
from 40 per cent in 1950 to 4 per cent in 2001 while agricultural tariffs have remained over 40 per cent. These sombre
facts remind us that ending hunger is a far more complex
problem than dealing just with population or food, a trap
into which well-intentioned single-interest groups are
prone to fall.

Exponential population growth continues, however.
Techno-optimists point to the fact that science and

technology have transformed the quality of life for many people and will continue to do so. But catastrophists warn of the danger of complacency because the dimensions of today's challenges are without precedent in the history of humankind.

Discussions about hunger polarise people into what Vaclav Smil of the University of Manitoba calls the cornucopians and catastrophists. Tim Dyson of the London School of Economics is quietly optimistic that farmers will be able to meet the volume of demand, provided there is no unforeseen calamity. Fertiliser use will increase, global cereal yields will reach about four tons per hectare, and information-intensive management procedures, including satellite remote-sensing and precision farming, will improve the efficiency of nitrogen application, soil quality and water usage. The big science question is how to invent a future in which food security is achievable for all and in an environmentally sustainable way.

Food security is defined as the secure ownership of, or access to, resources, assets and income-earning activities to offset risks, ease shocks and meet contingencies. In other words, not everyone is intended to be a subsistence farmer but everyone must possess the means to acquire an adequate diet. Today's farmers produce enough food to feed everyone in the world, yet 800 million remain food-insecure with food of insufficient quantity and quality for growth, activity and good health. If the world's supply of food had been evenly distributed in 1994, it would have provided an adequate diet of about 2,350 calories per day for 6.4 billion people – more than the actual world population. Physical redistribution of food in an equitable manner has so far proved impracticable and economically non-viable. It fails to supply the diversity of dietary needs and it does not generate enough income locally for small farmers – key factors for any attack on hunger.

Intensive systems of production and the green revolution have made major contributions to feeding the growing world population since the Second World War. Certain practices, however, are unsustainable and result in land loss through soil degradation, desertification and urbanisation, crop yield increases that have started to decline in less developed countries, and the irreversible depletion of biodiversity. Globally, only about 0.26 hectares of cultivated land is now available for each person to support food production and by 2050 it will be around 0.15 hectares. The rate of expansion of arable land has fallen below 0.2 per cent per year and continues to fall. About 40 per cent of the world's food requires irrigation and 70 per cent of the world's available water is used for agriculture. Raising food production to feed an extra two billion people will need to use the same amount of land with reduced amounts of available water.

Just as the world could not feed itself today with the farming methods of the 1940s, so farmers can hardly expect to meet the increased global demand in thirty or forty years' time with their present methods. Without another agricultural revolution based on more sustainable methods, the fate of the peoples of the less developed economies especially looks grim. India and China have the triple problem of expanding populations, declining water reserves and diminishing arable land, though the depletion in soil fertility in China is more regional than was once thought.

Will biotechnology help? History tells us that we should not expect too much from a single source. The second wave of products in the biotechnology pipeline could produce crops (including vegetables and fruits) that are engineered to give higher yields, longer storage possibilities and better nutritive properties. They could have resistance to certain pests and diseases or give high-value products such as vaccines. They may even contribute towards more sustainable systems of production if transgenic plants can be employed that require fewer chemical applications, provide better drought tolerance and grow in unfavourable or even hostile environments. In China alone, the area of transgenic crops has increased dramatically from less than 10,000 hectares in 1998 to 700,000 hectares in 2000 (44 million hectares worldwide). Yet the technology is too often perceived as the rich man's toy.

Nobel Peace Prize laureate, Norman Borlaug, father of the green revolution, has commented on the irony of farmers in the low-income, food-deficit nations being denied biotechnology when they have the greatest need of these new products. He recommends that transnational companies should share their expertise through partnerships with public research institutions, address agricultural problems not currently of high priority in the main transnational markets, and establish concessionary pricing structures in less developed countries so that poor farmers could benefit. Unless this happens, he believes hunger will remain largely a symptom of the failure of institutions, organisations and policies and not just the failure of markets and production.

Governments in poor countries have long discriminated against agriculture and been inclined to favour industry and cities rather than farmers and countryside. Less developed countries have been denied market access by more developed countries, aggravated by annual state payments to the agricultural sector by countries belonging to the Organisation for Economic Co-operation and Development that currently exceed Africa's entire GDP. If deepening poverty is to be reversed, the practical measures urgently needed include: applying the fruits of appropriate research and development, reforming land tenure, upgrading roads for marketing products, improving access to markets, and implementing good governance rather than corrupt practices and oppressive regimes. The European Union has gained valuable experience in strengthening disadvantaged areas of Europe and the time is ripe, foreshadowed by the Genoa meeting of G8 countries in 2001, for a comparable international effort in sub-Saharan Africa. Projects in India have already demonstrated how the communications and computer revolutions allow regions to leapfrog from outdated to modern technologies. Overcoming the prohibitive expense of protecting intellectual property is also a necessity. Bright young scientists returning home with new skills in plant biotechnology should be encouraged to find a central international fund through which they could protect their intellectual property. Such an initiative would send an important signal to the younger generation about the social, political and humanitarian intent to eliminate hunger in this century.

More than 1.3 billion people are absolutely poor with incomes of a dollar a day or less per person, and a further two billion are only marginally better off. We dare not allow this state to persist, because ending hunger is not an abstract human responsibility; it is central to sustainable development. If sustainable development is to be taken seriously, it will become obligatory for more developed countries to modify their consumption patterns for, as Nobel economist Amartya Sen has said, without change it will not be so much that humanity is trying to sustain the natural world, but that humanity is trying to sustain itself.

The emerging answer to the big science question is that we *can* end hunger but only if there is the political will to adopt appropriate policies and investment alongside the advances offered by research and development and appropriate technologies. Hunger will not end in regions where there is a continued lack of effective coordination of existing programmes of governmental and non-governmental agencies, let alone new initiatives. Failure to address such challenges plays into the hands of the catastrophists because it will perpetuate the denial of a basic human right for 800 million people – the right to an adequate diet.

Further reading

Tim Dyson: *Population and Food* (Routledge, 1996), 'World Food Trends and Prospects to 2025' (*Proceedings of the National Academy of Sciences*, USA, vol. 96, May 1999)

Brian Heap and Jennifer Kent (eds): *Towards Sustainable Consumption* (The Royal Society, 2000)

Christopher Leaver: 'Food for Thought' (28th Bawden Memorial Lecture to British Crop Protection Council Conference 2001, at www.bcpc.org)

Michael Lipton and Richard Longhurst: *New Seeds and Poor People* (Routledge, 1989)

Per Pinstrup-Andersen, Rajul Pandya-Lorch and Mark W. Rosegrant: 'World Food Prospects, Critical Issues for the Early 21st Century' (IFPRI, October 1999)

Jules Pretty: *The Living Land* (Earthscan, London, 1998); Amartya Sen: *An Essay on Entitlement and Deprivation* (Oxford University Press, 1981), Transgenic Plants and World Agriculture (Royal Society Report, July 2000)

Vaclav Smil: *Feeding the World: A Challenge for the Twenty-First Century* (MIT Press, 2001)

Anthony Trewavas and Christopher Leaver: 'Opposition to GM Crops: Science or Politics?' (EMBO reports vol. 2, no. 6, 2001)

Are We Still Evolving**?**

? It is tempting to suppose that in this modern era the human species has grown up and left its slightly embarrassing evolutionary history far behind. So what if we share 98 per cent of our DNA with the chimpanzee, our obvious evolutionary cousin? With modern science and technology, haven't we entered into an altogether new phase of history? Doesn't that 98 per cent point to the distant past, while that special 2 per cent – the part that accounts for that big human brain – points boldly to the future? Hasn't the blade of Darwinian evolution been dulled by our immense capacity to learn and adapt?

Between 1700 and 1900, life expectancy in Britain soared from seventeen to fifty-two years, mostly because of better nutrition and hygiene and cleaner air and drinking water. That, of course, was just the prelude to the feats of modern medical science. During the twentieth century, the death rate from infectious diseases in the United States fell by a factor of many thousands, and in 1977 smallpox was eradicated worldwide. We have artificial limbs, transfusions, pacemakers and heart transplants, and soon, perhaps, artificial tissues and organs. Scientists have mapped out the human genome, as well as those of the microbes that would threaten us. As Barry Bloom of the Harvard School of Public Health has noted, 'every gene of every human pathogen can be displayed in the computer screen of every student and researcher'. This, there can be little doubt, will multiply our adaptive skills still further.

From the biological perspective, organisms evolve by natural selection, as described by Charles Darwin. 'Fitter' individuals tend to live longer and have more offspring, thereby sending more of their genes into the next generation, while the genes of the less fit tend to get weeded out. But what if medical science gets in the way? Science has largely rendered a harsh world benign, strengthened the naturally weak and 'artificially' levelled the playing field of human reproduction. Does this mean that we have stopped evolving?

No – although the details depend on what we mean by 'we'. In the past decade, researchers have discovered that Tibetan villagers living at altitudes of over 4,000 metres are genetically adapted to do so. At least one extra gene helps their blood cells bind more oxygen, and also appears to boost their reproductive fitness under low-oxygen conditions. Within the past 10,000 years or so, these Tibetans have evolved genetically to be adapted to their environment. Many must have passed by the way because they were unable to cope with it. This is hardly surprising. Over the past twenty-five years, 1,173 people have successfully scaled Mount Everest. Some used extra oxygen, while others relied on that available in the free air. Many died during the subsequent descent, but it is telling that the chances of survival were nearly three times as high for those using supplementary oxygen. It is easy to imagine how the Tibetans may have evolved – even quite recently.

In the developing world, where some 25 per cent of all deaths still result from infectious diseases such as tuberculosis, pneumonia, Aids, malaria and measles, it is also clear that the population remains at war with the environment, as people's genetic defences battle against the genetically tuned weaponry of countless microbes. Recent research, published in *Nature*, has shown that in South Africa, for example, people with some genotypes will have a 30 per cent higher reproduction rate than others, because they will tend to live slightly longer through their peak reproductive years. There can be no suspicion here that evolution has come to an end.

Only if 'we' refers to the developed West does it become interesting to wonder whether evolution is over. Geneticist Steve Jones of University College London argues that modern medicine has largely taken natural selection out of the picture because people are living longer and dying differently. In the US, more than 50 per cent of all deaths now result from heart disease and cancer, which tend to strike later in life and after the reproductive years. Indeed, the fraction dying of heart disease has increased fourfold since 1900 – mostly because people are no longer succumbing to infectious disease. Jones argues that evolution might, if not have stopped, at least be slowing down.

Other scientists, however, suggest that not only are we in the West still evolving, but that we are doing so even faster than before. Biologist Christopher Wills of the University of California at San Diego suggests that climate change and other alterations to the environment, such as the ozone hole, are putting new environmental pressures on mankind. Moreover, more international travel means rising immigration and the mixing of gene pools that for millennia have been far more isolated.

Anthropologist Meredith Small of Cornell University argues that humans have not really changed the rules of natural selection at all. 'We might think that because we have culture – and with it all kinds of medical interventions and technologies – that we are immune from natural selection,' she says. 'But nature proceeds as usual . . . Some people live and some people die, and some people pass on more genes than others.'

Hence, by the definition of evolution, we are indeed still evolving. What may be more interesting, however, is how we are evolving – and in particular how our culture and science are now affecting this evolution. If you work with your hands, your skin gets tougher. This is not evolution, but adaptation – one of the simplest. The bluehead wrasse, *Thalassoma bifasciatum*, is a coral reef fish that lives in large schools made up mostly of females. There are at any time only a few dominant males. If one of these is removed, then one of the females will change sex and become a male, restoring the proper ratio. But the genes of this fish remain the same. It is just adaptation of a rather more sophisticated kind.

Human culture, likewise, is not stored in the genes – it resides in social structures and habits, in language and in the library, and has to be taught by one generation to the next. Culture, science included, is a learned adaptation of our society, and so, strictly speaking, not a direct consequence of evolution (although it is a consequence of our larger brains). The historian E.H. Carr pointed out the grave misunderstanding that can result from 'confusing biological inheritance, which is the source of evolution, with social acquisition, which is the source of progress in history'. That acquisition can be in the form of financial wealth, learning or what have you. In any case, it affects the future in an irreversible way, while leaving the genes untouched – at least for a time.

This argument was given a different spin by British zoologist Richard Dawkins in *The Selfish Gene*, published in 1976. He developed the idea of 'memes', cultural replicators or 'units of imitation', such as 'tunes, ideas, catchphrases and fashions', passed on from person to person in a similar way to genes. The idea divided the academic world, with some, such as philosopher Daniel Dennett and psychologist Susan Blackmore, enthusiastically adopting the idea as a key evolutionary influence, while others were dismissive. Palaeontologist Stephen Jay Gould described it as 'a meaningless metaphor' while biologist Steven Rose called it absurd to imagine 'the seamless web of culture being disaggregated and transmitted between minds by gene-like replication'.

But however it achieves it, culture undoubtedly does affect biological evolution. What are the myriad forces that make some people have children and others not? Disentangling the influences is surely very difficult, but these cultural effects have far more rapid consequences on the genetics of the human population than do the fairly rare genetic defects that lie behind many diseases. Small points out that at the national level one of the general consequences of economic and technological development is a marked decrease in birth rate. Regions with the highest birth rates are now Latin America, Africa and Asia, and these populations are therefore contributing the bulk of the genes to the gene pool of the human future. 'Culture may not seem a "natural" force,' says Small, 'but because it is part of our environment, it is just as natural as disease, weather or food resources.'

In the distant past, of course, it is certain that evolutionary forces favoured those with greater mental capacity, who could devise better tools and make better decisions. Wills argues that, within the human species, there is little reason to suspect that this has changed. On the other hand, he says, it could be that higher intellectual capacity tends to be correlated with an awareness of problems such as overpopulation, and could lead to a decrease in the average number of offspring of the really intelligent. If so, then evolution would be working in the opposite direction.

Where is the future leading us? The final bastion of biological evolution is the gamete – the sexual cell. No mutation or other manipulation of the physical organism gets carried on to its offspring unless encoded in the DNA of the gamete. How long with this remain true? When will some variant of gene therapy allow conscious manipulation of the DNA that is the gene pool? When this happens, and it is only a matter of time, then for the human species the distinction between biological and cultural inheritance will be over. Even then, of course, we will still be evolving.

Mark Buchanan
Physicist and science writer

Are we still evolving?

Michael Ruse
*Professor in the departments of zoology
and philosophy at the University of Guelph*

As Charles Darwin himself realised and discussed somewhat gloomily in his work *The Descent of Man*, there is more to evolution than natural selection. There has to be raw variation, on which selection can work. If everything and everyone were identical, then there could be no differential reproduction, and evolution would indeed grind to a halt. Although Darwin did not have much idea about the causes and nature of this variation – today we think it comes through mutation, which ultimately goes back to random changes in the DNA – he realised that if you stop or relax natural selection, evolution will continue (or perhaps start up again). If new variations are constantly coming into populations of organisms, and there is no selection either directing or removing them, then very quickly these variations in themselves will cause overall changes.

Darwin's big worry was that, thanks to modern medicine, many people with new or inherited deleterious variations would survive and reproduce, whereas before they would have died as children and failed to pass on these unfortunate features. As a decent caring person, he did not want us to stop helping the sick and ailing – and as a good Victorian libertarian, he would have been appalled at the idea that we might coerce people into foregoing reproduction – but he did point out that we would never let such bad practices persist with our domestic animals and plants.

I think Darwin was right in his science – protecting people from the forces of selection simply means evolution goes in other directions – but I am not as worried as he was about the consequences. I use beta blockers for high blood pressure, but they are now just as much a part of my environment as were those popping-with-grease, high-cholesterol, school-lunch sausages of my youth. Just as human culture has changed the forces of evolution, so human culture can protect us from the bad side effects of these changes.

However, this is not all there is to be said about human evolution today. While it may be true that we can shield ourselves somewhat from the forces of selection, it is hardly the case that we can do it completely – even in the developed world. And natural selection can and does occur, with full force, in the developing world. For example, it is evident among Aids sufferers in Africa, where some people just have a greater natural immunity to the HIV virus. It will be their offspring who survive and reproduce in greater numbers.

Another major ongoing evolutionary change comes thanks to modern travel and education and subsequent social mingling. Differences between races – differences of colour and size and (more controversially) qualities such as intelligence – are starting to break down through inter-breeding. According to the recent United States census, the children of Orientals and of Occidentals are increasingly rushing into unions with each other. What else would you expect at the University of California at Irvine, for example, where you have 30,000 bright, clean, intelligent, healthy, sexually active young people, away from home – 60 per cent of them of Asian origin and 40 per cent of European origin? If, a thousand years from now, we are all patterned on the prototype of Tiger Woods, who is the child of many races, I would be neither surprised, nor, given his magnificent physique, beauty and charm, much perturbed.

These are what you might call relatively natural evolutionary processes – at least, unplanned evolutionary processes. But since Darwin, eugenicists have dreamed that humans might take control of evolution and mould it for better ends. Of course, there is considerable doubt as to what constitutes a 'better end'. And, after Nazi Germany's dream of a race of superheroes, few now are willing openly to endorse deliberate racial breeding. However, certain kinds of biological population-shaping are occurring today and look set to increase. Take sex selection. In parts of the world – India notoriously – there is widespread use of techniques for discovering a foetus's sex and aborting if it is not the desired one. In practice, this almost inevitably means aborting if the foetus is not male. At a population level, this is bound to have significant effects on sex ratios (normally roughly fifty/fifty) and will have major trickle-down social effects.

The Human Genome Project, which has given us a map of the genetic nature of humankind, could potentially lead to a new, rapid and dramatic form of evolution, even if the conventional kind is fairly slow. Through manipulating our heredity, it may be possible to bypass conventional methods of reproduction and simply to design the kind of traits we want for our successors. At a minimum, we could try jiggling around with our existing frames, just extending or perfecting existing desirable features. Most people of my generation will live longer than our parents' generation because of better nutrition and related environmental and cultural changes, such as quitting smoking. Perhaps we might aim for a biologically based greater longevity too.

Yet, even something as modest as this could have strange side effects. If we live to, say, an average of one hundred years rather than an average of seventy-five, then I for one would hope that this would be accompanied by a little adjustment of the workings of the lower back. I do not welcome the prospect of an extra twenty-five years of back pain. How is this to be remedied or avoided? At the least, we would need to be stockier – bigger muscles, stronger skeletal frame – and perhaps leaning forward somewhat, so that the weight did not fall directly on the crucial bones and connections. But then we would need a longer and upwardly twisted neck, to avoid forever looking downwards. And probably it would be as well to shorten the legs, and to bend them forward a little, to maintain balance. In short, fixing the back would probably turn us into something that looks like an escapee from *The Lord of the Rings*.

Perhaps, after all, a short life and an upright one has things to be said in its favour.

While we have not yet gone this far, evolution – both natural and planned – is certainly still happening. But it has now moved from the biological to the cultural. We have not grown feathers, but we can fly. We have not developed bigger mathematical brain modules, but we can do calculations at far greater rates than ever before. We have no bigger talons or teeth, but we can kill by the millions. Here is the real point of human evolution today.

As we have evolved over the past four or five million years from ape-like creatures, the remains of ever more sophisticated artefacts show clearly that we humans have developed a more powerful method of change than anything offered by mere biology. In the physical world, changes start with random, undirected variations. Every time you get something good or worthwhile – a new efficient form or adaptation – it has to be produced anew in each generation, with the possibility that it will be simply diluted away by the usual processes of reproduction. It is a slow and inefficient process, at best. With the coming of culture – thinking, speech, artefacts, social systems, rituals – a new and advantageous idea can be passed straight on to everyone in the group almost at once, and then others can work to improve it and to put it into practice.

Problem-solving can be an art or a science or a craft and not just chance, and we can change and adapt almost at will. Suppose, for instance, that it would be advantageous to grow in height a few inches. Biology working through selection on variation would take many years to effect such a change. Culture can probe causes, using solution-seeking intelligence, and when the answers are discovered – a shot of hormone or improved diet – the information can be passed on at once to others without delay. This is why the answer to whether we are still evolving is yes. This is why humans, rather than our close relatives chimpanzees, flood the earth.

That said, biological evolution is crafty and conservative. After three and three-quarter billion years of ongoing change, nature has learned that it cannot afford to be seduced irretrievably by the latest great breakthrough in the selective process. One should not assume that culture is all-conquering. Instead, it succeeds best when it works with biology rather than against it. Consider those weird perversions of Christianity in nineteenth-century America. The Shakers proscribed sexual intercourse, and now they are down to two or three aged members, their sect chiefly remembered for its beautiful furniture. The Mormons stressed marriage and children and they now fill a full state of the Union, and a great deal more. In the early decades of the last century, some kibbutzniks decided that all children would be raised communally, with biological parents having no more or less contact with their own children than with those of others. Before long, everyone was cheating in order to steal opportunities to be with their own children exclusively. They were primates first and socialist pioneers second.

The point is that we are evolving culturally, but that this does not necessarily translate down into biological evolution immediately, nor can cultural evolution go as fast as it likes in whatever direction it likes. Evolution is a blind, thoughtless process and those who speak of the 'wisdom of the genes' do so metaphorically. It is a mistake to conclude that any tampering with what nature has produced is wrong. But natural selection is an ongoing and relentless force, and successful organisms – and we humans are certainly successful – work as well as they do because their parts function harmoniously together. Change has to be gradual and relatively smooth. There is good reason to think that our evolution is still in action, in part outside our control, in part, inside it. Which part is which, and whether we have the power to move the parts around at will, is another question.

Further reading

Daniel Dennett: *Darwin's Dangerous Idea* (Allen
    Lane, 1995)

Michael Ruse: *Monad to Man: the Concept of Progress in Evolutionary Biology* (Harvard University Press, 1997), *The Darwinian Revolution: Science Red in Tooth and Claw* (University of Chicago Press, 1999), *Mystery of Mysteries: Is Evolution a Social Construction* (Harvard University Press, 2001)
Christopher Wills: *Children of Prometheus* (Perseus Press, 1999).

Is There Life on Other Planets ?

? Definitive evidence of life on Mars. That was the confident claim made in early 2001 by Imre Friedmann of the American space agency NASA's Ames Research Center. It followed new studies of a Martian meteorite known as ALH84001. But not everyone shares Dr Friedmann's confidence.

What he and his colleagues have shown is that ALH84001 contains chains of crystals of a material called magnetite. Certain bacteria living on Earth have similar crystalline chains which, because magnetite is weakly magnetic, they can use to orient themselves within the mud in which they live. The ALH84001 chains have characteristics that, say microbiologists, point to a biological origin.

On this the critics agree. The question, they say, is when and where the chains got into the meteorite. ALH84001 has spent the past 13,000 years lying where it landed in Antarctica. Perhaps it became contaminated with earthly bacteria during that time. Friedmann refutes this. The chains, he says, were sealed in material that could only have come from Mars, and bacteria of the kind that use magnetite are not found in the Antarctic. The argument continues.

For anyone who likes simple answers to straight questions, this dispute is depressing. If scientists cannot even decide whether life once existed so close to us in our own solar system, how will they agree if there is or was life elsewhere in the galaxy and beyond?

The astrobiological approach followed by Friedmann et al. relies on searching for direct evidence of living systems. But it is not the only route. The best-known alternative is SETI, an acronym derived from the Search for Extra-Terrestrial Intelligence. The SETI Institute reasons that if life does exist elsewhere, it may have become at least as advanced as we are. If so, it will have developed radio transmission technology. Just as earthly broadcasters are inadvertently beaming *The Archers* into far infinity, so civilisations elsewhere must be disseminating their equivalent of our soap operas. SETI researchers therefore scan the electromagnetic spectrum looking for non-random signals arriving from space. Thus far, the search has gone unrewarded.

With so little evidence, it is not surprising that the question of extra-solar life has been dominated by theory and speculation. The universe is made up of billions of galaxies. Our own – the one we affectionately call the Milky Way – itself comprises more than 100 billion stars. Putting to one side any religious claims that a deity has selected our particular planet as the only place fit for life, there is a case on grounds of sheer probability for suggesting that humans are not alone.

In his classic science-fiction novel *The Black Cloud*, Fred Hoyle envisaged intelligence forming within a highly organised cloud of gas. But organisms as tangible as us are still the most likely possibility. Terrestrial life relies on the extraordinary capacity of carbon atoms to combine and recombine with themselves and with other elements in a myriad different ways. Not many atoms share carbon's versatility, so it would not be surprising if life even relied on chemistry similar to our own.

Life elsewhere would also need an environment able to sustain it – neither too hot nor too cold. It therefore seems likely that the bodies most likely to support life would be objects similar to Earth: solid, 'temperate' planets in orbit around a star. We know that planets exist because we see them in our own solar system. And in the past decade or so we have found direct evidence of many more.

In fact, astronomers now claim to have identified upwards of fifty planets orbiting stars other than our Sun. Most, like Jupiter, are large and gaseous and probably would not support life. But speaking at the annual meeting of the American Association for the Advancement of Science in 2001, Norman Murray of the University of Toronto presented new evidence of the existence of large numbers of solid, more Earth-like planets in our galaxy.

In their sample of more than four hundred stars, Murray and his colleagues reckon that more than half might have their equivalent of Earth. This is not to say that life would necessarily have appeared on them. But Murray is prepared to say that life could be common in the galaxy.

Some scientists go much further. They believe it is not just possible but highly likely that life has appeared elsewhere. Assuming that the behaviour of matter and the laws of physics are universal, they argue that the emergence of a system of molecular organisation of the kind we call life is inevitable. Others go further still: if such a process starts, the forces which resulted in evolution by natural selection here on Earth would inevitably produce intelligent life. But how often?

In 1961, with a confidence verging on hubris, the American astronomer Frank Drake devised an equation for calculating the number of technological civilisations in our galaxy. The Drake equation can be expressed thus: $N = R \times fp \times ne \times fl \times fi \times fc \times L$.

N is the figure he was trying to calculate – the number of civilisations in the galaxy that have developed to the point of being able to communicate. R is the rate at which suitable stars are formed – suitable, for these purposes, means likely to form planets. The next term, fp, represents the proportion of stars with planets, while ne is the number of planets round any star with a habitable temperature range. The three f factors represent the proportions of planets on which life evolves (fl), reaches the stage of intelligence (fi) and develops a communications technology (fc). Finally, there is L, the length of time for which an intelligent civilisation can hope to survive either accidental destruction by outside forces or self-destruction through misuse of its own technology.

The SETI people's attempt at fitting numbers into the equation puts the rate of star formation at about twenty per year. With a rising degree of arbitrariness, they suggest that half of all stars will form planetary systems, that the number of planets within one system that could support life is one, and that on one in five such planets life will appear and evolve. Mindful that whales and dolphins are intelligent but have never developed technology, they suggest that technology might be expected to appear in half the other worlds that support life.

Putting these numbers into the equation, you get: $N = 20 \times 0.5 \times 1 \times 0.2 \times 0.5 \times L$. That is, $N = L$. Or, to spell it out, the number of civilisations in the galaxy is equal to the number of years (L) that an advanced technological civilisation can hope to endure. The only such civilisation we have to go on is, of course, our own, which has only been seriously technologically advanced for some fifty years. So, the number of advanced life forms in our galaxy is fifty – at least.

This, of course, is just the SETI Institute's calculation. With assumption piled upon assumption, the Drake equation can be used to generate almost any result you like. However, Monica Grady of the British Natural History Museum, an authority on meteorites and the evidence of primitive life within them, says that scientists do take the equation seriously. 'I do think it still has a legitimacy,' she says. 'It is setting the framework, the likelihood of an extra-terrestrial civilisation. It gives us at least a back-of-the-envelope context.'

She reckons the chances of life elsewhere are about fifty/fifty, and like others draws a sharp distinction between any life and

Big Questions in Science

intelligent life; the former could have appeared time and again without progressing to the latter. Of the search for *intelligent* life in particular, many scientists are sceptical. Astrophysicist Michael Rowan-Robinson of Imperial College, London, points out that all planets have a finite lifespan. Sooner or later the stars on which they rely for their energy supply will grow dim (still several billion years to go in our case). A very long-lived civilisation would have developed the technology to do more or less anything it pleased, so responding to the imminent death of its star would surely become the number-one global project. Such a civilisation would be keen to communicate with, if not to colonise, a new world. In other words, we should have had some inkling of its presence.

'People who are firm believers in the existence of intelligent life elsewhere tend to go all mystical at this point,' says Rowan-Robinson. 'They say things like, "Ah, but they hide from us, they're able to communicate without us knowing." I just don't find this convincing.'

He accepts the argument that many civilisations reaching a certain point of development might tend to destroy themselves. 'But you'd think that any civilisation that had been around a really long time would want to leave behind a monument,' he says. 'A beacon sending signals targeted at people like us. Of course it may be that when you're about to blow yourself up you don't have time to think of erecting your equivalent of the pyramids! All the same, I'm surprised that nothing's been found.'

Should scientists ever discover firm evidence that intelligent life elsewhere does exist, it would not put an end to their search, of course. It would be just the beginning.

Geoff Watts
Writer and broadcaster on science and medicine

Is there life on other planets?

Colin Pillinger

Professor of planetary sciences at the Open University and leader of the Beagle 2 project, a part of the European Space Agency's Mars Express mission

? No one knows whether there is life on other planets, but new information gleaned from Martian meteorites makes it worthwhile having another look. It has certainly made it worth all the efforts my team has been making to obtain funds for the Beagle 2 project, which will fly to Mars in 2003 in an attempt to answer the question.

The subject has been good value for at least two hundred years; the first time it appeared in our daily ration of news was in the 1780s. The earliest story was not, however, from a science correspondent, but arose in a legal report of a court case: the defence counsel, in a trial for attempted murder, decided to make the case that his client was insane – and therefore not responsible for his actions – because he believed that men lived on the Sun. (It was dangerous ground because at the time the King's favourite astronomer, Sir William Herschel, believed that both the Sun and the Moon supported life.)

The lawyer was never required to prove his point – not because the defendant was obviously a very disturbed man (he starved himself to death while in prison), but because the judge had already ruled that the prosecution's case was flawed. There had been plenty of witnesses to events, the attacker was relieved of his weapon at the scene, and the intended victim had burns on her stays because the pistol was so close to her when it was fired. But in spite of all this, the learned judge nevertheless decreed that unless the ball the pistol had fired was produced, there was no evidence of intent to kill. After all, a man's life was at stake. And so it is with science: the more spectacular the claim you want to make, the better your evidence has to be.

Fortunately, science, unlike the law, is able to go on searching, almost indefinitely if necessary, for the bullet; we do not have to prove our case first time round. So far as life on Mars is concerned, we appear regularly to change our minds.

In 1976, the Viking mission, launched by the American space agency NASA, sent two spacecraft to the red planet with the best experiments it was possible to conceive to look for viable life forms. Very soon the results began to be beamed back. It appeared that something in the soil was utilising a radioactive material that had been sent on the mission as a possible nutrient source for any Martian organism that might be present. The mission also carried carbon dioxide with a radioactive carbon atom present, and some of this labelled gas was 'fixed' by an unidentified agent. In addition, when the samples scooped up were moistened, copious amounts of gas, mostly oxygen, were evolved. At least two, if not all three, of these results could be taken as evidence of biological processing.

Editors of the day should have been commissioning their stories and writing their headlines, but something held them back. A further experiment, carried out by the equipment on the spacecraft and designed to measure the abundance of organic material in the Martian soil, had failed to detect anything above a few simple molecules – and even these could be ascribed to cleaning procedures. The Viking scientists were faced with a dilemma. They had what looked like a biological function without the physical manifestation of its cause. Viking seemed to have life without a body. There was nothing for it but to opt for a non-proven verdict.

In fact, the scientists went a little further; they claimed that Mars was playing an enormous practical joke on them. Its surface environment showed signs of being incredibly oxidising so they explained away the results as chemistry mimicking biology.

Everybody fully expected that other space missions would rapidly follow Viking and that samples of Mars would soon be returned to Earth to demonstrate the truth or otherwise of the hypotheses formulated in 1976. But it was not to be. The non-proven verdict became not guilty, indeed it became the conventional wisdom that Mars was too hostile an environment to be considered a habitat for life.

For some scientists investigating other lines of enquiry, however, the case was never closed. Enter Martian meteorites. As a direct result of the Viking mission, it has been shown that we already have rocks from Mars here on Earth, thrown by giant impacts on the red planet. Painstakingly (every new piece of information has been extensively challenged) it has been revealed in a number of ways that these rocks have witnessed the passage of the essential ingredient for life – water – at temperatures appropriate for biological activity to occur. Organic matter, a chemical relict of biology, was discovered accompanying carbonate, a mineral deposited from water. Carbonate-containing sediments on Earth are petroleum source rocks – not that this means oil on Mars – but oil is one of Earth's most obvious manifestations of a vigorous past biology.

The carbon isotopic characteristic of the organics and carbonate on Mars can be shown to be very different; isotopic differences between co-occurring organics and carbonates are taken as demonstrating that biological activity began on Earth four billion years ago – in other words, almost as soon as the planet became solid enough to support a biology of sorts. This alone suggests that life might be rather easy to get started, not, as some would have it, some unique happenstance peculiar to Earth.

Then there was the case of the 'Martian fossil' – what appeared to be a nanometre-sized segmented worm. Love it or hate it, the Martian fossil, which, real or not, remains extremely controversial, has prompted the search on Earth for the smallest features in the geological record. It has made researchers look hard for corroborating evidence, such as the existence of biominerals – magnetite, for example.

So where are we now? Although much of what has been learned about Mars from Martian meteorites can be

Big Questions in Science

definitely demonstrated to have occurred on our neighbour in the solar system, a few important things cannot be proved. Most importantly, we cannot unequivocally demonstrate the provenance of the organic matter. This is why Beagle 2 is returning to Mars in 2003 armed with new experiments.

We are going to have a much more careful search for the bullet, to use the court analogy – the body that was so elusive for Viking. Beagle 2 carries a method based on combustion that detects every atom of carbon in all its forms. We are going to look for organic matter – chemical fossils – under the surface, particularly under a large boulder that has not been moved from the time it was deposited. We will also look inside rocks. The premise behind seeking out subsurface and interior locations is that organic matter would have a better chance of surviving protected from an oxidising environment.

Beagle 2's mass spectrometer will also be capable of searching for trace constituents of a biological nature in the atmosphere. The composition of Earth's atmosphere would be incredibly boring if it were not for the products of biology. All manner of species, perhaps the best known being methane, are there only because biology continuously produces them. It would be the same on Mars. If we could show that the oxidising atmosphere contained unstable reduced molecules, such as methane, which is the end product of the simplest metabolic reaction, we could argue for a biological source. That source might be 1,000 kilometres away or 1,000 metres below the surface but would nevertheless have to exist.

About 80 per cent of the population would like to believe that life does exist on other planets. We do not like being on our own. In my opinion, it is incredibly arrogant to think that humankind is the pinnacle of evolution. While we probably will not have a definitive answer about planets beyond the solar system for many years to come, it is worth bearing in mind that in science the absence of evidence is not evidence of absence.

Further reading
Frank Drake and Dava Sobel: *Is Anybody out There?* (Delta, 1994)
Seth Shostak, *Sharing the Universe* (Berkeley Hills Books, 1994).

Big Questions in Science

How Will the World End ?

? At the opening of the sixth seal, there will be a devastating earthquake, the Sun will blacken and stars will fall from the sky. Mountains will shake. At the blowing of the fifth trumpet, locusts from the bottomless pit, with human faces and scorpion tails, will savagely torment the unfaithful. Seven plagues will cause calamity upon calamity, including loathsome sores, rivers turned to blood and a heat so severe that the world will be scorched.

The Bible's Book of Revelation contains all the vital ingredients for an end to the world that is satisfying to the human psyche. Many civilisations have believed that the end will be an essentially human-centred affair, a purposeful event in which evil is punished through some divine and apocalyptic means, and redemption follows in a new world for those who deserve it.

These visions of doom remain with us, but today there is also a new kind of doomsday prophet who is often atheistic and scientific. While there are members of this breed who are as fanatical as any traditional doomsday merchant, there are also those who are deeply pragmatic about their cause. Their argument is that we are at a unique point in human history because we face a host of new scenarios by which the world might end – either the whole world or the human world as we know it. These endings have nothing to do with religious belief or divine intervention; they are real and their likelihood can be calculated, measured and, in some cases, avoided.

Some of the endings these new prophets describe are uncannily reminiscent of age-old themes. An asteroid crashes into Earth, bringing a familiarly cataclysmic end to the human race. Wanton exploitation of nature brings about humanity's fall, demonstrating that humankind contains within it the decadent seeds of its own decline. Yet there is something materially different about many of the claims about the end over the past fifty years: the rise of science.

Science has enabled us to understand the natural world better and therefore to imagine and predict many previously unthought-of apocalypses. It has also enabled us to devise means by which we could, accidentally or deliberately, finish off much of the human race, and perhaps the planet.

Paul Corcoran, associate professor at the University of Adelaide, argues that this change became obvious some fifty years ago with the explosion of the atomic bomb. 'The end was indeed nigh and it required no belief in ancient prophets or the intervention of deities to accept it as the actual condition of one's life,' he writes in *Awaiting Apocalypse*. 'This was an apocalyptic vision suddenly reduced to practical choice and the probabilities of chance error ... the active contemplation of The End has been for all intents and purposes both a rational and an ordinary intellectual and emotional experience for half a century.'

The twenty-first century is a crucial century for the survival of our species, according to Martin Rees, Britain's Astronomer Royal. This is because we are acquiring the means by which to extinguish ourselves but have not yet devised a way of spreading ourselves through the galaxy, a move that would enhance our chances of survival by reducing our dependence on a single planet. The new risks we face fall into several categories, perhaps the most gruesome of which is the deliberate obliteration of humankind by humankind. Nuclear war was but the first possibility. Some argue that global annihilation through war is more likely to be the result of bioweapons – cheap to produce, easy to conceal and terrifyingly tricky to control. The World Medical Association, which represents about eight million doctors, warns that the consequences of a successful biological attack, especially if the infection were readily communicable, could far exceed those of a chemical or even a nuclear event.

We have been promised for decades that the end will come at the hand of robots, generally acting under the command of some malevolent inventor, and subsequently rebelling and taking over the Earth. Scientists such as Hans Moravec, one of the founders of the robotics department at Carnegie Mellon University, predict that, despite delays, every year machines are getting nearer to becoming conscious. Once they become brighter than us they could dominate us, wipe us out or merge with us in some sort of post-human synthesis that represents the end of humanity as we know it.

Several of the new dangers to the world resonate to a time-honoured theme – that of decline and fall brought about by moral bankruptcy. The latest incarnation of this idea is environmental degradation. John McNeill, in *Something New Under the Sun: An Environmental History of the Twentieth Century*, argues that the twentieth century was unique because of the scale on which this became possible.

Human curiosity is also traditionally viewed as a prelude to doom. Scientists have always lived with metaphors (Pandora's Box,

Frankenstein) that illuminate the way in which unseemly curiosity and meddling with nature, however innocent their motives, could destroy the world. Modern incarnations of these destructive forces include genetically engineered microbes. These could bring about the end of the world either through environmental disaster by breeding superweeds – a possibility discussed by American economist and environmental campaigner Jeremy Rifkin – or by their use in manufacturing bioweapons.

The end could also come as the result of the blue-sky tinkerings of physicists. Pottering away at their particle accelerators with the aim of understanding abstract physical questions of the universe, scientists could trigger a chain reaction that would destroy the world. In 1983, Piet Hut and Martin Rees suggested in an article in the journal *Nature* that the Relativistic Heavy Ion Collider (RHIC) on Long Island, New York, could create a subatomic black hole that would slowly eat away at our planet. Alternatively, the RHIC might create exotic bits of altered matter, called strangelets, that would obliterate whatever ordinary matter they met.

Although a panel called to address these fears has now rejected both scenarios as virtually impossible, critics say that that is not enough. 'Totally impossible' is the only acceptable probability when it comes to the preservation of humanity.

Another branch of physics, nanotechnology, could also cause an exotic end. Engineers have for the last decade been building tiny, atomic-scale machines. One day they may be able to build microscopic robots that can assemble and replicate themselves. The technology, which could produce such benefits as performing surgery inside the body, could also become lethal and uncontrollable. They could, according to Eric Drexler, author of *Engines of Creation*, 'reduce the biosphere to dust in a matter of days'.

Before we get around to destroying ourselves, however, the apocalypse may arrive from outer space. Once it was divine omnipotence that made these types of utterly unpredictable ends all too believable. Now it is our understanding of the vastness of galactic phenomena and Earth's cosmic insignificance. Astronomer Duncan Steel suggests in *Rogue Asteroids and Doomsday Comets* that we could eventually be wiped out by an asteroid. Alternatively, Earth might be obliterated by a gamma ray burst – a sporadic explosion that produces far more energy than does our Sun. Earth's atmosphere would initially protect us from most of the burst's deadly X-rays and gamma rays. But they would slowly cook the atmosphere, a process that would destroy the ozone layer. Without the ozone layer, ultraviolet rays from the Sun would reach the Earth's surface, killing off the tiny photo-synthetic plankton in the ocean that underpin the world's food chain.

If the gamma ray bursts – which are rare – fail to get us, a rogue black hole just might. Scientists have calculated that our galaxy contains about 10 million black holes and they usually orbit stars, which means that none is likely to approach us. However, if one did just scrape through the solar system, it would exert enough gravita-tional effect to distort planetary orbits. Earth's orbit might become elliptical, causing extreme climate swings, or Earth might shoot out of the solar system to a bitter, cold end in outer space.

There is another stellar danger whose potency we have actually witnessed in the last decade, thanks to vivid images from spacecraft. Solar flares are magnetic outbursts that bombard Earth and can dis-rupt power supplies, though our atmosphere and magnetic field have so far afforded good protection. There is evidence, however, that ordi-nary Sun-like stars can occasionally emit superflares, millions of times more powerful than ordinary ones. Bradley Schaefer of Yale University has found evidence that some perfectly normal-looking, Sun-like stars can brighten briefly and massively, and believes that this may be because of superflares.

These could be vastly more worrying should Earth lose the pro-tection of its magnetic field. Geologists have shown that, every few hundred thousand years, this magnetic field dwindles almost to nothing for perhaps a century, then gradually reappears. The last such reversal was 780,000 years ago, so the next one may be due. One sign that it may be coming is that the strength of our magnetic field has decreased by about 5 per cent in the past century. Without protection from the magnetic field, Earth would be vulnerable to particle storms, cosmic rays from the Sun and further erosion of the ozone layer.

A particularly twentieth-century terror has been of the end at the hands of aliens. Today, not only does SETI (the Search for Extra-Terrestrial Intelligence) scan space for signs of intelligent life, but theologians and philosophers have prepared our moral response, should aliens ever be discovered. Cutting-edge thinkers on this issue, extrapolating from the age of human exploration of the globe, argue that the greatest danger from aliens is not that they will unkindly kill us all but that we may suffer because we get in the way of their plundering of Earth. Alternatively, they might import diseases to which we have no resistance. Finally, as foretold in the late Douglas Adams's *The Hitchhiker's Guide to the Galaxy*, they might simply wish to extinguish us and our homes as part of some greater scheme, such as the construction of an interstellar bypass.

Perhaps, however, divine intervention will precede all this. And if not mediated directly by God, it could be at the hands of some of his most dedicated followers. The End – decline, apocalypse and redemption – has a gripping effect on many, and these days small cults, such as the Branch Davidians and Heaven's Gate, who wish to accelerate its coming, have unprecedented access to potent methods of retribution. A taste of what they might achieve was given us in 1995, when members of a religious sect released nerve gas in a Tokyo station, killing twelve people and injuring more than 5,000. Since September 2001, we have been confronted with the concept of the global holy war, an ancient idea made more lethal through modern technology. With improved access to bioweapons, and even to nuclear bombs, we may get the ending with which humans began.

Aisling Irwin
Award-winning science writer

How will the world end?

John Leslie

Philosopher and fellow of the Royal Society of Canada

Our galaxy contains many billion Sun-like stars. Our telescopes see details of many billion further galaxies. Assuming that intelligent life evolves easily, why do we detect no signs of extra-terrestrials? Can it be that we humans are the very first intelligent beings to evolve in this neck of the cosmic woods? That there will, in due course, be a million other technologically advanced species, but ours is the first one? Surely this would make our position altogether too extraordinary. Wouldn't it be more sensible to think that, yes, many intelligent races have evolved before us, but they have then destroyed themselves soon after developing advanced technologies?

Here is another line of reasoning which looks almost exactly like it. Can you and I really be among the first millionth of all *humans* who will ever have been born? Wouldn't that make our position unbelievably extraordinary too? If all humans believed they were in the first millionth, only one in a million would be right. This point occurred to Brandon Carter, British cosmologist and mathematician, in the early 1980s. It led him to what has come to be called 'the doomsday argument' for thinking that humans will quite soon be extinct.

If looking simply at the various dangers confronting the human race, without taking account of Brandon Carter's point, then how high ought we to put the probability of humans surviving for millions of further years? At about the 80 per cent level, I'd answer. But what if it really were that high? And what if (as most space scientists think) a human race that survived for very much longer than it already has would spread right through its galaxy, perhaps in as little as 600,000 years? You and I would then presumably have lived among the earliest millionth of all humans. Doesn't that look too incredible?

Incredible, at any rate, when you consider the entirely credible alternative. There has been a population explosion recently. Of all humans who have lived so far, roughly 10 per cent are alive at this very moment. If it were a moment near the end of human population history, then our position in that history would be none too extraordinary – nothing like as extraordinary, at any rate, as being in the first millionth. And aren't there plenty of threats to the continued survival of our species?

While such reasoning is controversial, it seems to me to show that 'doom soon' quite probably awaits the human race. 'Soon' here means soon enough for galactic colonisation to have scarcely got started. And this could well mean doom during the next few centuries.

Let us examine just three of the ways in which our species could become extinct quickly. Two are widely discussed. They are extinction through a pollution crisis and extinction through biowarfare. The third may come as a surprise to you. It has been considered, however, in articles in physics journals and in *Before the Beginning*, a recent book by the British Astronomer Royal, Martin Rees. It is extinction through a vacuum metastability disaster.

A pollution crisis could involve many factors. The ozone layer, Earth's shield against ultraviolet radiation, could become severely eroded. There could also be a steady accumulation of poisonous chemicals. Land could become exhausted despite ever heavier application of fertilisers. Fertilisers could themselves be a threat, bringing death to rivers, lakes and even seas. Still worse, accumulating greenhouse gases might make temperatures shoot up disastrously.

In order to achieve the consensus needed for influencing politicians, the Intergovernmental Panel on Climate Change has paid scant attention to worst-case scenarios. In these scenarios, harmful changes produce additional changes of the same type. For instance, vegetation could die from excessive heat, and then land losing its vegetation would get hotter, leading to more death of vegetation, and therefore more heat, et cetera.

Again, vast amounts of methane, a greenhouse gas already approaching carbon dioxide in importance, could be released from warming tundra and from the continental shelf sediments of warming oceans. Eventually water vapour, at present tending to form clouds that reflect more heat than they trap, could come to be a greenhouse gas disastrous in its effects.

What would be absolutely the worst scenario for overheating produced by pollution? Famous for the hypothesis that natural checks and balances have until now kept our environment stable, independent scientist James Lovelock gave us his answer in his book *Gaia*. Earth, he said, could heat up 'to a temperature near to that of boiling water'.

Vacuum metastability is a danger much less known about than pollution, and may even be illusory. The idea is that the space we inhabit, 'the vacuum', is far from entirely empty. It is filled with scalar fields. These are characterised only by intensity, not by the directionality that makes a magnetic field detectable by compass needles. Fish on the ocean floor cannot know about water pressure, which is the same wherever they swim. Similarly, humans could be unaware of the scalar fields, their intensities remaining identical for as far out as telescopes can probe. But if they exist, which is what most physicists think, then they determine the properties of every atom. Change any such field, and those properties change. Change it in a way that reduces its potential energy, and the alteration spreads with enormous violence. Reduction of potential energy is what happens to a ball after it gets knocked out of a hollow on a hillside where it had sat 'metastably'. The ball gets to do what it wants to. It rolls downhill. Thanks to physicists investigating very high energy densities, the space around us, too, might get to do what it wanted, through being jolted out of its metastable condition.

We remain sadly ignorant about the physics of very

high energy densities. If vacuum metastability manifests itself here, then physicists might some day create the ultimate environmental catastrophe. They might produce a tiny bubble that immediately expanded at nearly the speed of light, destroying first our planet, then the solar system, then all the stars in our galaxy, then all the nearby galaxies, and so on.

This process could not begin tomorrow or even during the next decade. Experimenters would have to push beyond the energy densities known to be safe, which are the ones attained by colliding cosmic rays. At present, our particle accelerators are below those energies by a factor of perhaps several million. But, in his *Dreams of a Final Theory*, Steven Weinberg talks of using powerful laser beams to accelerate individual charged particles to the Planck energy – roughly equivalent, he explains, to 'the chemical energy in a full automobile gasoline tank'. Not even colliding cosmic rays pack anything like that amount of punch.

Maybe scalar fields are entirely fictitious. Possibly, they exist but are fully stable like a ball at the bottom of a valley. But if they exist and are only metastable, then not even beginning to colonise the galaxy would make the human race sure to survive. Picture those people in their spaceships, happily congratulating themselves on being safely away from an earth laid waste by pollution. Then they see the expanding bubble. A few seconds later, it overtakes them.

The physics of this area is so difficult that perhaps only actual experiments could prove that it held no risks. In a book section headed 'Premature Apocalypse?', Martin Rees writes that 'caution should surely be urged (if not enforced)'. Personally, I'd be all for enforcement.

The most likely cause of premature human extinction, however, seems to me biowarfare, rather than vacuum metastability or pollution – although the temptation to go to war could certainly be increased by a polluted, collapsing environment. So far, the most environmentally ravaged nations have been the ones least able to resort to force because they are too impoverished. But germs could become the poor man's atom bomb. While crops and live-stock could be targets, the main risk of human extinction comes from germs designed to attack humans. A single bottle filled with tiny beads can now yield viruses in quanti-ties that would earlier have come only from large factories. And advances in genetic engineering have made it easy to design utterly lethal organisms.

An aggressor nation's vaccination programmes or other measures to protect itself might fail, or a nation losing a war might be willing to take the terrible risk of killing everyone. A terrorist organisation might threaten to put an end to all humans, then actually carry out the threat when its demands were refused. Experiments dating from 1948 show that just a few aircraft distributing a disease such as smallpox in aerosol form could infect almost every-body in Britain.

The world's end might also come about by accident. Recently, a genetically modified mousepox killed every infected mouse without exception. It had been created by Australian researchers – by mistake!

Further reading

Paul Corcoran: *Awaiting Apocalypse* (Macmillan, 2000)

John Leslie: *The End of the World: The Science and Ethics of Human Extinction* (Routledge, 1996)

J.R. McNeill: *Something New Under the Sun: An Environmental History of the Twentieth Century* (Penguin, 2001)

Duncan Steel: *Rogue Asteroids and Doomsday Comets* (John Wiley & Sons, 1997).

What is Life About ?

? So here you are, a poor, forked being, thrown into the world, with death the main forthcoming event. What, exactly, is the point? Most of the answers seem to involve a story of some kind. What matters then is who tells the story, who the protagonists are and whether it has a moral.

The message from science for much of the twentieth century, and the cue for a good deal of existential angst, was that the moral was that there was no moral. It began with British analytic philosopher Bertrand Russell reacting, or overreacting, to news of the eventual heat death of the universe, by pronouncing sonorously that 'only on the firm foundation of unyielding despair can the soul's habitation henceforth be safely built'. Sixty years later, molecular biologist Jacques Monod was putting a scientist's gloss on the insistence by philosopher Jean-Paul Sartre that the universe was indifferent to human purposes. Monod depicted life as the product of 'pure chance, absolutely free, but blind'. So the typical response to a request for directions about the meaning of life has been 'I wouldn't start from here'.

Lately, though, there has been a revival of attempts to read meaning into scientific stories, even to establish a curious kind of secular religion based on an epic telling of the history of the universe. Just as postmodernism was supposed to have dissolved grand narratives, science has pieced together the grandest narrative of all, running over fifteen billion years and counting. And it is a narrative in which life plays a starring role.

But why should anyone think that science has anything to say about what life is about in the first place? The origins of that notion lie deep in the history of Western ways of making sense of the world. The Judeo-Christian tradition underwrote a fixed, hierarchical cosmos, ordered in the way American philosopher Arthur Lovejoy described in *The Great Chain of Being*. God was at the top, but in the earthly realm humans had dominion over everything else. And, of course, some humans had dominion over other humans.

Just as our modern struggles to rethink assumptions about race, sex and class are still marked by that tradition, so are our attitudes to ecology and the natural environment. In the 1970s, the historian and cultural critic Theodore Roszak attributed part of the modern malaise to what he dubbed the desacralisation of nature, his term for the removal of God to a separate realm. He identified this with the rise of Christian theism. If the universe was mere matter in motion – even though God might still be the prime mover – and other creatures Cartesian animal machines, then humans could basically do whatever they liked – with the rest of creation, if not necessarily with each other. Roszak suggests that if God falls out of the picture, all restraint is lost, hence the horrors of the twentieth century.

What humans liked took on a more earthbound focus with the coming of modernity. There might still be a heavenly paradise to come, but material progress was also worth striving for. And, as biologist Steven Rose argues, with the advent of evolutionary theory, progress came to seem a principle of the natural world as well as a social possibility. Philosopher Michael Ruse has documented in detail how even evolutionary theorists who say they are not progressivists still find it hard to do without the idea that somehow evolution is moving upward as well as onward. It was always, according to Ruse, 'a secular religion'.

E.O. Wilson, the American pioneer of sociobiology, which finds biological explanations for social behaviour, argued that evolution both explained why humans needed religion, and supplied the best religion achievable in a secular world. In the late 1970s, in *On Human Nature*, he told how scientific materialism 'presents the human mind with an alternative mythology that until now has always, point for point in zones of conflict, defeated traditional religion. Its narrative form is the epic, the evolution of the universe from the big bang.' Twenty years later, in *Consilience*, he came back to the idea that 'people need a sacred narrative . . . If the sacred narrative cannot be in the form of a religious cosmology, it will be taken from the material history of the human species.'

A number of writers have picked up Wilson's cue and tried to present the history of the universe, and especially the history of life, in religious terms. For biologist Ursula Goodenough in her meditation on the meaning of her discipline *The Sacred Depths of Nature*: 'The big bang, the formation of stars and planets, the advent of human consciousness and the resultant evolution of cultures – this is the story, the one story, that has the potential to unite us, because it happens to be true.'

Other scientists have invoked the fashionable theories of complexity and self-organisation to suggest that, in the phrase of theoretical biologist Stuart Kauffman, we are 'at home in the universe'. In his popular book of that title, he argues that many things we used to see as staggeringly unlikely are in fact expected outcomes of relatively simple rules. And this, he believes, is a radical departure from science's historical role as a threat to our self-esteem. Copernicus elbowed us out of the centre of the universe. Darwin showed how evolution was a glorified dice game over billions of years. Freud even persuaded us that we knew little of our own thoughts. But Kauffman's message is a counter to the modern view that we sit in an obscure corner of an immense universe, the end product of a history of natural selection, which, in Monod's phrase, amounts to no more than 'chance caught on the wing'.

Current scientific thinking seems to support the epic of evolution more than Kauffman's less strictly Darwinian version of the origins of order. In *Green Space, Green Time: The Way of Science*, Ursula Goodenough's fellow evolution enthusiast, the American science writer and environmental activist Connie Barlow, proposes rituals that celebrate 'the pageant of life' and the wonders of an evolutionary universe. She is looking for a big story that allows her to promote the ecological values that she believes are needed to overcome global environmental crisis. And, while she is inspired by Wilson, she also takes ideas from biologists such as Lynne Margulis, whose study of the 'microcosmos' of bacteria persuades her that cooperation, not competition, drives evolution. James Lovelock, Margulis's co-founder of Gaia theory – which suggests that the whole Earth functions as a single self-regulating system – is often depicted by these tellers of the universe story as a secular saint. Although Lovelock is known to be dubious about humanity's place in the great scheme of things, he is overflowing with warm feelings for the planet, rooted in his scientific vision. 'I find myself looking on the Earth itself as a place for worship, with all life as its congregation,' he writes. 'For me this is reason enough to do everything in my power to sustain a healthy planet.'

Look at contemporary science with a mind open to wonders, and it is easy to see how proponents of a new religious naturalism are enthused. Yet they have a problem. We can marvel at the scope of the evolutionary epic, feel our interconnectedness with all life (the various genome projects help here by showing just how much we have in common with other creatures) and strive for continuity. But we are cast as sophisticated spectators. Our role as conscious beings is to unravel the story, and celebrate it, to conserve what we can and value biodiversity for its own sake. The future beckons with infinite opportunities for bearing witness and, er, that's it.

Others see humans centre stage, as the beings smart enough to speed up the next leap in evolution. Californian futurist Gregory Stock, for example, in his popular book *Metaman*, says almost exactly the same thing as Barlow: 'We now know the basic outlines of a history of life and the cosmos so rich that it can serve as a powerful modern mythos to orient our lives and our vision of the world.' But his vision of the future is quite different. We are on the verge of a new evolutionary transition, so let's get to it. The global superorganism, a merging of technology, culture and the biosphere, is already visible in outline and our immediate destiny is to help it into being. The enlightenment project is alive and well and will be realised through a grand synthesis of information technology, biotechnology and global systems. Transcendence is still on the cards, but it is transcendence for materialists.

So both schools of thought have rejected the pessimistic answers to the question of what life is about. They mostly regard Jacques Monod as a 1970s curiosity. They have answers to the conclusion of theoretical physicist Steven Weinberg in *The First Three Minutes* that 'the more the universe seems comprehensible, the more it also seems pointless'. For them, the universe is a self-organising project for the evolution of complexity. But what if the question is: what is the point of a human life, or even all human life, in the context of that great unfolding? The fact that there are such radically different answers suggests that they are far from obvious deductions from the science. As usual, the story, however strongly rooted in the best science around, does not come with a ready-made moral. That has to be supplied in the telling.

Jon Turney
Head of department, science and technology studies, University College London

What is life about?

Steven Rose

Professor of biology at the Open University and joint professor of physic at Gresham College

What is life about? Well, it depends who you ask. Great religions and secular philosophy alike have always been more than willing to provide answers: to live virtuously so as to perform God's will and as a preparation for the next world; or to be part of the cycle of rebirths leading ultimately to Nirvana, for instance. To live to serve society, to enjoy life hedonistically, or to accept stoically, non-believing humanists of various stripes might respond. Or might life be full of sound and fury, signifying nothing, as the despairing Macbeth insisted?

Inevitably, all these classical answers interpret the question as relating to human life. Traditionally, since René Descartes in the seventeenth century, non-humans were regarded as mere machines whose life was not about anything; only humans, endowed with a soul, could have a life that had purpose. In the nineteenth century, two transformative biological discourses revolutionised both the questions and the answers. First, materialist physiology in the Continental European vein simply dropped the 'about' as having nothing to do with science. Teleology, purpose, the aboutness of things, was taboo, as it seemed to sneak some sort of mysticism or theism in by the back door. Instead, the task was to concentrate on the seemingly simpler but ultimately equally problematic question: what is life? Then, in England, Charles Darwin's evolutionary theory suggested that life was 'about' something for animals as well as humans, that animals were not merely there to serve humanity's needs but had purposes – teleonomies – of their own.

School biology texts in my youth adopted the

physiologist's approach and listed a series of characteristics of all living organisms. These varied, but tended to include metabolism, growth, repair, irritability (that is, the capacity to respond adaptively to environmental stimuli), homeostasis or self-regulation, and of course reproduction. By the 1950s, physicists, following Austrian physicist Erwin Schrödinger (but indeed with undertones going back to the Reverend William Paley with his argument from design for the existence of God), suggested that living creatures were characterised by organisation, by entropy. Their point was that we are all negentropy machines, resisting the otherwise universal tendency to disorder.

Such lists of characteristics raise problems. For instance, are viruses, incapable of independent reproduction outside a cellular host, alive or not? These definitional problems used to seem of more abstract than practical significance, but became a matter of impassioned – and costly – debate over the prospect of life on Mars, bedevilling the various space missions sent to detect its presence. How would one know it if one found it, barring little green aliens with antennae on the tops of their humanoid heads? And when we get to the 'artificial life' world of the computer buffs, the temperature rises several more degrees. Can such defining features of life be incarnated (inmachinated?) in silicon rather than carbon chemistry? Self-taught biologist Steve Grand caused a stir worldwide when he developed the computer game *Creatures*, which simulates the life of creatures called norns. These exist only on the computer screen but can within that environment be 'born', reproduce and 'die', and can react to their environment, including fellow norns. But can they be said to be alive? No NASA probe would pick them up, for sure.

Physiology, concerned with proximate mechanisms, has nothing further to say about the about question. Indeed, it is almost smug in its prim refusal to have anything to do with it. Instead, it built on the once fashionable science of cybernetics – the communication of information – to square the teleology circle, made hygienic by being called teleonomy instead. Rather than seek purpose in the world, we could at least characterise living creatures as showing goal-directed behaviour. We act purposively in no more or less a manner than does a room thermostat in regulating temperature around a goal or set point.

The newly powerful claim to purpose in life has mainly been left to the neo- (or fundamentalist) Darwinians, whose religious fervour on the question brooks no dissent. Behind physiology's proximate processes come what the fundamentalists call 'ultimate' explanations. Note the biblical undertone, which is far from accidental. For them, the purpose, the telos, of life is reproduction, the spreading of one's genes – that is, pieces of DNA similar or identical in sequence to pieces of the DNA in one's own cells – into the next generation. Organisms – the bodies that contain these bits of DNA – are mere vehicles designed by DNA to accomplish its reproductive goal. As British evolutionary biologist Richard Dawkins puts it, with his customary eloquence, 'elephant DNA is a gigantic program which says "Duplicate Me" by the roundabout route of building an elephant first'. In this view, an organism is a mere vehicle, the lumbering robot created by genes for the purpose of enabling them to replicate. In so far as the organism itself has a purpose, therefore, it is to aid this process, by producing the maximum number of offspring capable of surviving and reproducing in their turn (a property known as fitness). Or, as any organism has DNA sequences in common with those of its near relatives, it is to aid its siblings or cousins to reproduce (inclusive fitness).

This has the virtue of consistency and simplicity. It offers a coherent world view – an answer to all possible questions. In this godless age, it has come to serve religious purposes, as biologist Brian Goodwin and sociologist Dorothy Nelkin have observed. The ACG and T of DNA replace the Yahweh of my orthodox Jewish upbringing. It even enables one to speak of evolutionary ethics – the claim that ethical principles can be derived from evolutionary mechanisms and hence ultimately from DNA. Just as the Christian gospel makes the enigmatic claim that 'in the beginning was the Word' so this DNA fundamentalism locates the origin of life in the emergence from an abiotic alphabet soup of the predecessor of modern DNA, a naked replicator.

I reject this claim for life's origin and purpose by insisting that chickens come before eggs. Outside the metabolic cellular web in which it is embedded, DNA neither replicates nor has functional meaning. Bringing DNA to life requires metabolism, energy, cellular structure – in a word, organisation. It is noteworthy that the discussion over whether meteorites from Mars contained living organisms centred on whether the microstructures observed within them were bounded by the characteristic lipid membranes possessed by all living cells.

Putting chickens – cells and organisms – before eggs – DNA – reminds us also of another key feature of all living organisms. Although physiology textbooks continue to speak of homeostasis, using the room thermostat image, life is, of its nature, not static, but dynamic. Our body constituents are in a continuous state of flux, being synthesised and degraded every second of every day. Stasis is death. This led the great biochemist Frederick Gowland Hopkins to define life as 'the expression of a particular dynamic equilibrium which obtains in a polyphasic system'. We are all born, develop, age and die. Each of us humans, and indeed each and every other living creature, exists as a unique trajectory – or lifeline – through time and space.

It is this dynamic concept of development, of a lifeline – a concept which returns the organism, rather than the gene, to the centre stage of life – which provides the basis for my alternative answer to the 'about' question. For organisms are not merely the expression of a programme or blueprint unrolling from their genes, modulated as this may be by environmental contingency. Rather, organisms construct themselves – create their own trajectories – out of the raw materials provided by their genes and their multiple levels of environment, from the cellular to the social. This process of self-construction goes by a number of names. The philosopher Susan Oyama speaks of the 'ontogeny of information' and 'developmental systems theory'. The biologists Humberto Maturana and the late Francisco Varela coined the phrase 'autopoiesis' for this process of self-construction.

Whatever we call the process, it provides another way of approaching the question of what life might be about – a route that transcends both old determinisms and the studied rejectionism of physiology. Life is about both being and becoming; we are in a state of constant transformation. A newborn baby suckles; within a few months it has developed teeth and chews. Chewing is not simply suckling writ large; it involves different muscles, and different neural processes. So a baby has to be simultaneously a competent suckler, and becoming a competent chewer. Being and becoming, regulated by intricate self-organising control systems at multiple levels of organisation. In this sense, all organisms construct their own future, though in circumstances not of their own choosing. While this may be what *all* life is about, it is overwhelmingly true for humans, with our large brains and social organisation, and a sense, however partial, of our complex biosocial histories. It is these capacities which enable us to glimpse, in however dark a glass, what might be to come. And within these constrained necessities comes the freedom to choose, to act, and to build towards not just our own future but that of the whole of humanity and the planetary economies in which we are embedded.

Further reading

Connie Barlow: *Green Space, Green Time: The Way of Science* (Copernicus, 1997)
Ursula Goodenough: *The Sacred Depths of Nature* (Oxford University Press, 1998)
Jacques Monod: *Chance and Necessity: An Essay on the Natural Philosophy of Modern Biology* (Vintage, 1972)
Steven Rose: *Lifelines: Biology, Freedom, Determinism* (Penguin, 1998), (edited with

Big Questions in Science

Hilary Rose) *Alas Poor Darwin: Arguments against Evolutionary Psychology* (Jonathan Cape, 2000)

Jon Turney: 'Telling the Facts of Life: Cosmology and the Epic of Evolution' (*Science as Culture*, 10(2) 2001, pp. 225–47)

Steven Weinberg: *The First Three Minutes: A Modern View of the Origin of the Universe* (Basic Books, 1993).

Big Questions in Science

superstrings (M theory) 20
Swinburne, Richard 2, 4

taxi drivers 52
telescopes 12, 15, 24
Terman, Lewis 71
terrorist attacks 133
test-tube babies 144
thalidomide 144
Thermodynamics, Second Law of
    23–24
thought:
    animals and 50–51
    computers and 162
    models of 47–48, 49
    nature of 46–54
Thurstone, Louis 73
Tibetan villagers 188–89
time:
    age of 28
    big-bang and 27
    constants of nature and 28
    direction and 22
    energy and 26
    future of 30
    gravity and 26
    looking into past 12
    mass and 26
    nature of 21–30
    traditional view of 26
time travel 29–30

transcutaneous nerve stimulation
    167
Trewavas, Anthony 177
tuberculosis (TB) 155–56, 157
Twell, Henry 26
twins 71

ultimate explanations 228
United Nations Food and Agricul-
    tural Organization 176

United Nations World Food
    Summit 176
United States of America 180, 189,
    196
universe:
    background radiation x, 12, 15,
        16, 25
    beginning of 11–20
    cosmic repulsion 17
    earliest phase of 14, 15, 18, 22
    energy in 17–18, 18
    expansion 12–13, 16, 18, 29, 30
    future of 16–17
    homogeneity of x
    inflation 19, 25, 29
    mass 19
    organisation in 24
    standard model 14

vaccinations 155
vacuum metastability disaster 216,
    217–18
Vaneechoutte, Mario 84
Varela, Francisco 36, 230
vasopressin 132
Ventner, Craig 52
Victoria, Queen 166–67
Viking mission 205
Virchow, Rudolf 154
vitamin A 179
Voltaire 92

Wall, Patrick 166, 167
Wambugu, Florence 179
Warnock, Dame Mary 144
Warwick, Kevin 73
water 177, 182
Watson, J.B. 93, 155
Watts, J.C. 142
Webb, Sidney and Beatrice 93
Wedekind, Claus 122
Weinberg, Steven 5, 218, 225